Ground Subsidence

Ground Subsidence

A.C. WALTHAM
Department of Civil Engineering
Trent Polytechnic
Nottingham, UK

Blackie
Glasgow and London

Published in the USA by
Chapman and Hall
New York

Blackie & Son Limited,
Bishopbriggs, Glasgow G64 2NZ
and 7 Leicester Place, London WC2H 7BP

Published in the USA by
Chapman and Hall
a division of Routledge, Chapman and Hall, Inc.
29 West 35th Street, New York, NY 10001–2291

British Library Cataloguing in Publication Data

Waltham, A.C. (Antony Clive), 1942–
Ground subsidence
1. Soils. Subsidence. Geotechnical
aspects
I. Title
624.1'5136

ISBN 0-216-92500-2

For the USA, International Standard Book Number is

0-412-01801-2

Photosetting by Thomson Press (India) Limited, New Delhi
Printed in Great Britain by Bell & Bain (Glasgow) Ltd

Preface

This book arose through a longstanding interest in limestone subsidence and collapse, which subsequently evolved into a study of all forms of ground subsidence. Most information on ground subsidence has previously been available only in disseminated form within various technical journals, and so this new book is the first to present a comprehensive overview of all aspects of ground subsidence.

Each style of subsidence is considered in its own right, because the causative mechanisms vary enormously. For example, sinkholes and collapses over random natural voids clearly contrast with the widespread and inevitable man-induced subsidence of drained peat; thus each subsidence process merits its own chapter. Mining subsidence, in its various forms, is covered together with the many other ways in which man can induce subsidence in his own environment. Subsidence on clay is discussed without getting too deeply involved in the intricacies of load-induced building settlement. This book has been structured so that for each subsidence process, fundamental mechanisms are assessed, and illustrated with case histories; these are followed by critical reviews of appropriate site investigation techniques, hazard recognition and an appraisal of remedial engineering work.

The book is directed towards practising geotechnical ground and civil engineers, and to final year undergraduates and postgraduates in engineering geology, civil engineering and related fields. Specifically, it has been written by a geologist who has become aware of the requirements of the engineering profession through many years of working with and teaching civil engineers. Geologists and geomorphologists may find within the book some insight into the wider applications of their own subjects.

Finally, the author is happy to record his thanks to Blackie and Son for their encouragement and guidance, to Peter Smart for much constructive advice, to Trent Polytechnic for providing facilities, to Margaret Fry for excellent typing, to numerous colleagues and friends for discussions and ideas, and to his wife Jan for her endless support.

ACW

Contents

1 Introduction

Among the major environmental hazards, ground subsidence is often left as the poor relation to landslides, earthquakes, volcanoes and floods. Perhaps this is because individual cases of subsidence are usually not as dramatic as a volcanic eruption or a major landslide, and only rarely involve loss of life. Yet subsidence may rank as the most widespread ground hazard. It is not restricted to narrow earthquake belts or steep slopes, but can occur in widely scattered situations, normally where the geological causes behind the potential failures are hidden below the ground surface. In terms of ground engineering, subsidence is among the most insidious of threats, and therefore warrants careful assessment at the site investigation stage, especially in certain geological environments.

Styles of ground subsidence vary greatly. Slow long-term movements, perhaps due to sediment dewatering or some methods of modern mining, can disturb large areas, with gentle but ultimately destructive effects. Alternatively, instantaneous ground collapses, such as those where soils drop into buried limestone caves or abandoned mines, can cause total havoc over small areas of ground, but are usually separated by long time intervals of relative stability. Some cases achieve a wide notoriety. The Leaning Tower of Pisa, the Florida sinkhole collapses, the disappearance of the peatlands and the drowning of Venice are all well known, but each has been caused by a separate suite of geological processes and events. More threatening than any of these is the suite of processes (referred to in Chapters 8, 10 and 11), creating the widespread subsidence in the Central Valley of California, which has already incurred enormous costs and could ultimately prove very expensive indeed to the valuable agricultural industry of the Valley (Prokopovich, 1986c).

One way of classifying ground subsidence is to start with a division into endogenic, natural processes and exogenic, man-made processes (Prokopovich, 1986a; Costa and Baker, 1981), but this provides a very uneven split. The purely natural processes, including tectonic movements, volcanic deflation, rock solution and deep sediment compaction, are almost entirely very slow and of limited impact on most engineering time scales. In contrast, man's activities can induce far more rapid and serious subsidences, notably through

Figure 1.1 A bus caught at an unusual angle when a road collapsed beneath it early in 1988 in the city of Norwich. The subsidence was due to failure of weathered chalk above old abandoned mines (Photo: Eastern Counties Newspapers).

mining or fluid withdrawal—the latter causing clay consolidation, peat wastage, accelerated rock solution and soil collapses. In all these cases, the subsidence is the result of interaction between man's activities and existing geological environments. The aim of this book is to consider the range of geological processes which lie behind the ground subsidence. Consequently, the only subsidence classification employed is based on the geological parameters, and has determined the headings of the chapters following.

The spectrum of subsidence processes, variously influenced by man and nature, creates problems in delimiting the subject area, and building settlement lies astride the boundary. Its style and extent are largely the result of the imposed ground loading, and it already has its own extensive literature within the fields of soil mechanics and foundation engineering. Consequently it is reviewed only briefly in Chapter 9, by way of introduction to some of the more severe cases of settlement where the geological conditions are especially relevant to the scale of ground failure. Subsidence in areas of permafrost is not referred to in this book, as it too has little geological background and is almost entirely due to man's disturbance and melting of the ground ice beneath and around his own structures; the subject has its own literature, with numerous case histories (such as de Ruiter, 1984) and useful reviews by Ferrians et al. (1969) and Andersland (1987).

The literature on ground subsidence lacks any previous book which draws together all its different styles and aspects. There are some wide-ranging review papers (including Allen, 1969; Carbognin, 1983; Coates, 1983) and others more relevant to individual areas (including Prokopovich, 1972; Malkin and Wood, 1972). Some volumes of collected papers delve widely into the various styles of subsidence (Holzer, 1984; Johnson et al. 1986; Saxena, 1979), while others provide valuable coverage within narrower subject areas, including fluid abstraction (Poland, 1984), sinkholes (Beck and Wilson, 1987) and mine workings (Forde et al., 1984). Among the literature on ground engineering, the volume edited by Bell (1987a) provides a comprehensive overview, and offers a link with the world of the practising civil engineer who is usually in the front line in dealing with ground subsidence.

Perhaps the underlying message from any of these reviews or the chapters that follow, is that of the major role played by man in inducing the subsidence of his environment. Natural ground subsidence processes have changed little over the millennia, but the escalating worldwide pressure on the resources of the land mean that ground subsidence is having an increasing impact in areas of development and urbanization.

2 Cavern collapse

Natural voids can occur in a wide variety of geological situations, though their numbers are dominated by solution cavities in limestones. They comprise the aspect of geology perhaps least understood by many engineers, and site descriptions with sweeping statements about the distribution of concealed voids are often grossly misleading. Except in certain areas, notably of limestone, natural cavities are not often encountered, so the ignorance concerning them persists, but where they do occur they can create hazardous ground conditions which are among the most difficult to predict or remedy (Culshaw and Waltham, 1987).

Natural macrocavities rarely occupy more than 1% of a rock mass, though they can be localized into some dangerously cavernous horizons or fracture zones. Even there, they rarely achieve the void ratios produced by some styles of mining, but they make up for this by being unpredictably erratic in their distribution. Cavities can be formed by various types of underground mechanical erosion, or alternatively by fissure opening in the head zones of landslides. However, by far the most important is underground chemical erosion; solutional processes are the only ones which can achieve results in the initial stages of development, when weaknesses in the solid rock are limited to microfractures too narrow for mechanical transport. Furthermore, solution is most effective at producing large cavities where its effort is concentrated along well-defined fracture lines and not diffused through the rock mass: hence the importance of cavities in massive limestones.

Limestone caves

Most caves are formed in strong, massive limestones, with low rock permeability, which contain widely spaced fractures available for solutional enlargement by through-flowing groundwater. The same applies to most dolomites and marbles. Weaker, more porous limestones have more diffuse groundwater flow and rarely develop large cavities; caves are rare, but not unknown, in chalk (see Chapter 4). Tufa and travertine may contain caves at shallow depth where deposition by surface streams encloses voids beneath the projecting shelves of waterfalls.

Caves are diagnostic components of karst landscapes—which are in turn characterized by the underground drainage through the caves. Though developed in a variety of styles, most karst terrains are easily identified by their sinkholes, dry valleys, closed basins and rocky landscapes (as further described by Palmer, 1984, and Jennings, 1985). Limestone is such a common rock that a worldwide overview (Middleton and Waltham, 1986) shows few countries without karst and caves, though many limestone regions form the less developed uplands. The largest areas of limestone, with the greatest impact on engineering practice, are in southern China and the southeastern USA.

Limestone solution rates are directly related to the amounts of rainfall, and also to the levels of groundwater carbon dioxide, most of which is derived from plant and bacterial activity within the soil cover (White, 1984). Consequently the luxuriant vegetation of the wet tropics, in places such as Malaysia and south China, is optimal for cave development; limestones in these environments are normally more cavernous than similar lithologies in the cooler climatic zones. Nevertheless, natural solution rates are so low that around 5000 years is required to form a limestone cave a metre in diameter (Mylroie and Carew, 1986). The engineering hazard on limestone is not due to cave growth, but is entirely due to the failure of old caves, or, far more commonly, the failure of soils into old limestone caves to form sinkholes (see Chapter 3).

Limestone caves exhibit infinite variety in size and shape (Ford, 1988). They may be networks of narrow fissures following the pattern of rock fractures, or rambling mazes of spacious tunnels (Figure 2.1). Isolated solution cavities may be less than a metre across, while the largest known cavern, in Sarawak, is 700 m long and 400 m across; this has an impressively low, arched roof rising only 100 m, but as it is comprised of unusually massive limestone 300 m thick, there is no imminent danger of collapse. Many cave roofs are stable arches or cantilevers in sound rock, but flat roofs spanning wide caverns in thin-bedded

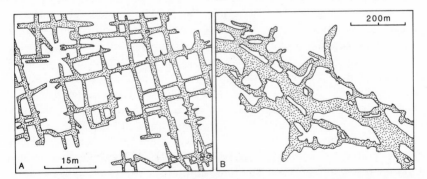

Figure 2.1 Plan views of two contrasting styles and scales of limestone cave development. *A.* Small joint-controlled maze cave in the English Pennines; nearly all the openings are along the two sets of vertical joints within the horizontal limestone. *B.* Large cave complex in Sarawak; there is no recognizable pattern within the warren of very large cave passages. In both examples, many of the passage ends are just choked with sediment, so the limestone voids are even more extensive than mapped.

Figure 2.2 A cave chamber in the strong Carboniferous limestone of the Yorkshire Pennines with no signs of recent collapse. Even though the rock is well fractured, blocks on the left stand out in cantilever, and the flat roof span is stable with the surface less than 10 m above.

limestone may exhibit upward migration by progressive failure. Caves can occur at depths up to 1000 m, but 10 m of solid rock cover provides a stable roof to all but the larger caverns, making the deeper caves irrelevant to most engineering works. A bank in a small town in America has its underground vault only 6 m above the ceiling of a 25 m wide river cave; though the cave was unknown at the time of construction, there is, fortunately, no danger of collapse as the limestone is strong and, in this case, unusually little fractured.

Caves are commonly controlled by bedding, shale interbeds, joints and faults, and the geological influences are often recognizable on cave maps (Figures 2.1 A, 2.4). However, the high density of potential control features, and their random selection by the cave development, makes predictions of cave locations impossible. Caves do not lend themselves to statistical analysis, and the only feature normally predictable about caves is that they are unpredictable. Investigations for the Keban Dam, in Turkey, revealed a random collection of cavities (Bozovic *et al.*, 1981), and boreholes in the Kuala Lumpur limestone, in Malaysia, find multilevel caves (Tan, 1987) with no recognizable pattern. Boreholes in the Hershey Valley of Pennsylvania found voids in 11–17% of the core length (Foose and Humphreville, 1979), but all the sampling was in the top 10 m of highly weathered limestone beneath a pinnacled rockhead. It is only rarely that caves exceed a few percent of the

Figure 2.3 Progressive upward collapse of the roof of a large cave in the Carboniferous limestone of South Wales. The original, solutionally-eroded cave passage was more than 10 m below the level now seen, but the thin beds of limestone have progressively peeled away from the roof to form the huge pile of floor debris. This has occurred over a timespan of thousands of years, and, as this cave is over 100 m below the surface, there is no ground hazard at this site.

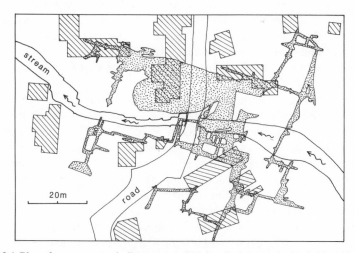

Figure 2.4 Plan of cave passages in limestone underlying the houses of a Yorkshire village. The narrow cave passages are mostly joint-controlled fissures, but the largest cavern on the map is a low, wide chamber eroded from a nearly horizontal bedding plane. The only entry to the cave is in the left bank of the stream just below the bridge. There is almost no collapse hazard as most of the cave is more than 15 m below the surface.

volume of a limestone mass, but they can still constitute significant engineering hazards.

One justifiable comment on cave distribution is that the density of caves is likely to be locally higher adjacent to an impermeable rock outcrop; there, surface drainage provides concentrated flows of unsaturated water into the limestone and maximizes solutional effort. Also, major faults and, in some areas, particular beds of limestone can sometimes be recognized as being more susceptible to cave development. Other than their entrances, which may be open only by chance, caves normally have no surface expression, making ground assessment always difficult.

Limestone cave collapse

Limestone collapse events are extremely rare due to the long time scales of solutional void enlargement. Of the thousands of sinkholes in limestone terrains (see Chapter 3), very few involve rock failure. In Florida's sinkhole areas, Sowers (1975) records just one rock collapse forming a vertical shaft 20 m deep, and perhaps one other rock failure where a number of piers collapsed instantly beneath a bridge, though some cave entrances are prehistoric collapses. Similarly, soil collapse sinkholes dominate in China, but some modern rock collapses are recorded (Zhang, 1984), and a sinkhole 176 m deep in Sichuan almost certainly involved some rock failure.

The Sichuan event approaches the size of massive prehistoric collapse features which are now part of the landscape in various countries including Mexico, Sarawak and New Guinea; examples in Yugoslavia are illustrated by Božičević and Pepeonik (1987). Hundreds of metres deep and wide, these caves may have developed progressively, enlarging over a long period as the cave rivers undercut the walls around the growing piles of collapse debris (Quinlan et al., 1983). As examples of engineering hazards, they are statistically ignorable.

More significant are small caves encountered during foundation work, such as that into which a dozer fell on a Saudi construction site when a thin cave roof collapsed beneath it (Grosch et al., 1987). Caves only become a threat within the zone of construction loading, and so can normally be ignored at depths greater than 30 m. Mangan (1985) illustrates caves intersected on French construction sites; the variety of their morphologies demonstrates the unpredictability of their occurrence, and begs the question of what other caves lie concealed beneath these and other sites. Any calculation of rock roof stability over a cave is difficult, as conventional engineering approaches are conservative, and attempts at evaluation have found natural spans survive with about three times the span of artifical openings at the point of collapse (Cruden et al., 1981). As a rough guide, a limestone cave roof is generally stable where the thickness of sound rock cover exceeds the cave width. Less cover may present a failure hazard, but will then depend largely

Figure 2.5 Blocks of limestone, tilted and fallen where they have subsided by a metre or two due to the roof collapse of a low, wide bedding-controlled cave in the Yorkshire Pennines. Collapse zones like this are not common and represent a low level of subsidence hazard, but old collapse areas buried by soil can present difficult foundation conditions.

on the fracture patterns within the limestone, and each cave must be assessed individually in its local context.

Caves in other rocks

Gypsum and salt are widespread rocks even more soluble than limestone, but low strength and high solution rates make progressive ground subsidence (see Chapter 7) dominant over cavity collapse. Salt is the extreme case, and caves are recorded only in a few arid areas, such as Israel's Mount Sedom. Gypsum is the intermediate case, and substantial caves can form in it; the most extensive gypsum caves are in the Russian Ukraine where one area of just 65 ha is underlain by 105 km of cave passages, each a few metres high and wide, at depths of 20–50 m, but modern collapses appear to be unrecorded.

Next to the soluble rocks, the greatest numbers of caves are found in basalt lavas. Formed by progressive solidification of a surface lava crust while the molten lava still flows beneath, lava tubes or lava caves are normal features of pahoehoe flows, as their underground conduits are the only means by which lavas can flow large distances without excessive cooling (Wood, 1981). The caves are mostly 2–15 m high and wide, and individual tubes may be many kilometres long. Due to their origins, they all lie at depths of only a few metres; their roofs are often little more than the solidified lava crust, and the frequent natural collapses form the only entrances (Figure 2.6). Fortunately they are found almost only in lavas of Holocene age (< 10 000 years), as older caves have collapsed, been eroded away or been infilled; but in young volcanic areas, such as Hawaii and Iceland, they do form an engineering hazard. In Iceland a

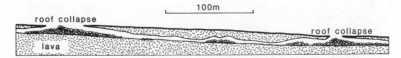

Figure 2.6 Profile through part of a lava cave in Hawaii formed in a 1973 lava flow. There has been some subsequent roof breakdown, which in two places collapsed through to the surface around the time of a small earthquake (after Wood, 1981).

main road stands on only 4 m of rock above the 10 m wide lava tube of Raufarholshellir. The cave entrance lies only 60 m from the road, and is itself a collapse, but the cave was not mapped till after the road was built; the cave roof is progressively collapsing, though monitoring with nets across the cave floor shows the failure rate is erratic but slow, and creates no immediate hazard.

Quartzitic sandstones in southern Venezuela contain some huge caves at depth with collapse holes reaching 200 m wide and deep; they appear to have developed by hydrothermal solution and subsequent surface-related erosion. A cave system contains 186 000 m^3 of passages and caverns beneath a 40-ha outcrop of dipping calcareous conglomerate near Krasnoyarsk in Siberia. Although these examples demonstrate the potential for eroded caves in other rocks, they appear to be rarities not yet encountered in engineering works.

Much more widespread are soil pipes (reviewed by Jones, 1981). These are mostly 5–50 cm in diameter and occur where there is point-to-point through-flow of water in a suitable soil; this is commonly a fine soil with high dispersion, often rich in smectite and prone to fissuring, with cohesion low enough to permit internal erosion but high enough to bridge the resultant voids. Pipes may also develop in loess, especially where thin sand horizons provide the initial drainage route; the Kansas loess has some caves up to a metre in diameter, but they collapse and run in to form depressions 25 m across (Landes, 1933). Terrace edges and gully walls are sites where piping is propagated by steep groundwater gradients, and limited areal subsidence may occur, on a long time scale, due to extensive piping beneath a stronger caprock (Jones, 1981). Pipes also form in soils which drain into underlying fissured limestone, and this is the main process of sinkhole formation (see Chapter 3). Vertical pipes may reach considerable diameters before collapsing to create the sinkholes. In addition, lateral pipes over 2 m in diameter may extend 20 m or more, and Newton (1987) cites examples in Alabama which led to subsequent ground collapse.

A type of cave that can form in any rock is the landslip fissure, often misleadingly known as a tectonic cave. These develop under tension in the head zones of sliding slabs of competent rock. Interbed sliding may leave fissures with a rock roof, though most extend to the surface and are then filled with soil, when a soil bridge may subsequently develop; open cavities are rarely more than a metre wide. The commonest type is the gull, formed along a camber fold: a competent rock fractures and sags, downward and outward, along a scarp edge or valley side, as an underlying clay is squeezed out due to

Figure 2.7 One of the collapse entrances to the Hawaiian lava cave shown in Figure 2.6.

Figure 2.8 The rubble slope in an Hawaiian lava cave below a collapse zone where the roof has failed by upward stoping and is now within a few metres of the surface.

the loading (Figure 2.10). In England, gulls are widespread along and above slopes in nearly all sandstone–clay or limestone–clay sequences of Carboniferous age or younger. The Windypits are well-known examples in the limestone

Figure 2.9 Gulls exposed in a road cutting through the cambered escarpment of the Magnesian Limestone in Nottinghamshire. The gull on the right is over 30 cm wide and was partly filled with soil and fallen rubble from the weathered zone, but the soil cover is continuous and there is no surface expression of the void.

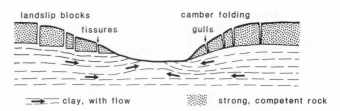

Figure 2.10 Cross-section through a valley with gulls and fissures developed in the sides by camber folding and block landsliding over a weak clay which has been squeezed out and then eroded away from the valley floor. The strong caprock is most commonly either a sandstone or a limestone, and the gulls and fissures are often covered or partly filled by soil.

of the Yorkshire Moors; with only small surface entries, open fissures reach depths of 40 m and lengths of more than 100 m (Cooper *et al.*, 1982).

Comparable fissures in level ground may develop by ice-drag where moving Pleistocene glaciers created tension in competent bedrock which could slide above a weak bedding horizon. Hundreds of metres of cave up to 3 m high and wide have formed in this style and now lie beneath a housing area in Montreal (Schroeder *et al.*, 1986), but more over-run slabs of limestone form roofs 5 m thick and have minimized the collapse hazard.

Site investigaton

Due to their random, unpredictable distribution, natural cavities are notoriously difficult and expensive to assess and locate in site investigation. Certain geological environments, as described above, and notably any known cavernous limestones adjacent to impermeable outcrop, can be identified as hazardous; but the geological controls cannot be interpreted adequately for prediction for engineering requirements. Geophysical surveys may offer some guidelines (see below) but cavity searches must then resort to drilling. The exception is a search for gulls, where trenches dug perpendicular to the scarp edge or valley side prove more useful (as in Hawkins and Privett, 1981).

Simple probing, or percussion drilling, is perfectly adequate for most cavity searches, and costs are around a tenth those of rotary-cored boreholes. The drilling penetration rate clearly distinguishes rock from voids or soft cave fills, and loss of flushing medium also indicates cavernous ground.

The density of probes needed to find small, but still hazardous, voids is far in excess of that used in normal site investigation. The limestones under Kuala Lumpur, Malaysia, normally merit ten times the borehole density used in other rocks, often with 100 holes drilled on the site for a single high-rise building (Tan, 1987). A common practice is to place a probe below the site of every intended structural pile. The Remouchamps Viaduct in Belgium has five piers on limestone, and 4–8 boreholes were drilled at each pier site in the initial investigation (Waltham *et al.*, 1986). Only very small voids were found by these few boreholes, but large caves were then revealed in excavations for two of the pier foundations (Figure 2.11). Subsequently 308 probes were drilled on the five sites, but revealed no more cavities. Though this number of holes does represent some 'overkill', the operation revealed the frustrating unpredictability of cave distribution.

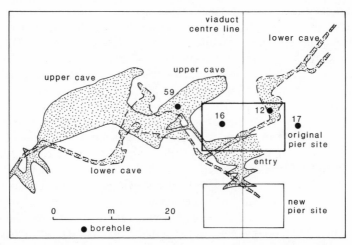

Figure 2.11 Plan of caves and exploration boreholes at one of the pier sites for the Remouchamps Viaduct in Belgium. The cave was found only when excavations at the original pier site broke through the soil and collapsed rock which buried the entrance. Boreholes 12 and 59 had ended above the cave. Subsequently part of the upper cave was filled with concrete, and the pier was moved to a site away from the main cave; the lower cave is 25 m below the ground at the pier sites.

—The optimum depth for exploratory probes is also open to debate, though a common practice is to regard 5 m of continuous sound rock as adequate reason to terminate. However, deeper holes at the Remouchamps Viaduct site (Figure 2.11) managed to miss all the caves, and one hole 15 m deep stopped just above a cavern. On a single site it may be possible to recognize more cavernous zones, either in certain limestone beds or at distinct elevations (notably in the wet tropics, as in Tan and Batchelor, 1981), and it would be false economy to stop probes even in solid rock just above these.

Once voids have been found on a construction site, more probes may be needed to delimit cavity sizes, or a downhole camera may prove useful. Continuous reassessment may involve only a small time penalty and allow a more economic investigation than an initial plan for a vast number of holes; but any adequate exploration of cavernous ground is forced to utilize a relatively high density of bored holes. Entering and directly mapping the caves is the alternative and more efficient technique, but is frequently not possible.

Geophysical detection of caves

The geophysical techniques which may be applied to cave detection have an extensive literature, reviewed at length by McCann et al. (1982, 1987), Bates (1973), Owen (1983) and Rat (1977) and more briefly by Greenfield (1977, 1979). Sadly, the overall conclusion is that little of geophysics is reliable enough for widespread use in this role for engineering purposes, but the high potential cost savings from a reduced borehole search means that some applications do warrant consideration.

Perhaps the most useful technique is the gravity survey, as a cave will obviously generate a negative anomaly. A 2 m diameter void beneath a 1 m thick rock cover creates a 42 μgal anomaly, while the same under 9 m of rock registers only 10 μgal; the anomalies are roughly halved if the cave is water-filled. Surveys to this level require a modern, very sensitive gravimeter, and still encounter the problem that background noise is typically around 30 μgal. Microgravity surveys were used to trace the Lost River Cave beneath Bowling Green, Kentucky (Crawford, 1986); one section, located by a traverse with a 3 m station interval, was verified by drilling and entry into a cave 12 m wide, 2 m high and 10 m down. Current work at Western Kentucky University is finding that 7 m diameter caves at depths of 15 m produce anomalies around 100 μgal; however, similar anomalies have been drilled to reveal only low-density collapse zones, and the same has been recorded by Greenfield (1979), though these still represent hazardous ground. The vertical gradient of gravity is close to 300 μgal/m, but decreases to 200 μgal/m over caves. Measurement of this, using a wooden bi-level tower, has proved easier and more successful than Bouguer anomaly surveys (Smith and Smith, 1987), and has found caves 3 m across and 15 m deep in the Florida limestone.

Electrical surveys mostly rely on recognition of induced ground resistivity anomalies, which, relative to limestone, are positive through caves but negative through clay-filled cavities. Most recorded profiles, using conventional traversing arrays over known cavities, produce data which in all honesty cannot confidently be related to the caves, as anomalies can be confused with background noise and are often offset. However, the pole–dipole method (Bristow, 1966) has achieved repeated success even though it is based on the questionable premise of hemispherical equipotential surfaces; its system of identifying overlapping anomalies helps penetrate background noise, and with multiple readings it generally provides useful data, though it has completely failed on some tests over known caves. As modified by Bates (1973), the Bristow method has found caves 4 m across at depths of 12 m in Alabama (Owen, 1983), and it appears to be the most useful electrical technique, though only in areas of simple geological structure.

Among seismic techniques, refraction surveys do not identify voids whose depth is greater than their size. Reflection surveys are also of little value, though some limited successes have been recorded (Kirk and Snyder, 1977). Resonance surveys have had some claims to success but are poorly understood, and have been tried and prematurely abandoned due to interpretation difficulties on a road project over cavernous ground (Arrowsmith and Rankilor, 1981). Cross-hole seismic surveys have some value where enough holes are available to define the site profile and hence recognize anomalies (McCann et al., 1982).

Ground-probing radar is seriously limited by its depth penetration, though it has found small caves to depths of 5 m (Bjelm et al., 1983), and has also proved useful in cross-hole surveys (Ballard et al., 1983). Of the passive techniques, magnetic surveys suffer from the lack of contrast between limestone and air, but are potentially useful over shallow lava tubes. Ground electromagnetic emissions which could relate to buried voids remain unverified, and ground heat-flow, detectable by airborne infrared photography, may relate to shallow caves and their soil drainage impact (Greenfield, 1979) but has so far failed to achieve useful reliability.

In conclusion, reliable detection of small cavities, which can still threaten structural performance, is impossible by any geophysical technique, except perhaps the radar survey with its severe depth limitations. Interpretation of any method is hampered by background noise, notably from rockhead irregularities, and many attempted surveys have failed to produce useful results. But the high cost of extensive borehole programmes can still make any geophysical guidelines economically welcome. Particularly in certain limestone terrains where large caves are known to occur, geophysical surveys, of which gravity surveys are perhaps best, may usefully identify ground anomalies. These always need back-up drilling, but their recognition can greatly improve the efficiency of a borehole exploration.

Foundations on cavernous ground

The variations in shapes and sizes of natural cavities are such that generalizations concerning ground treatment are rarely appropriate even on a single site, and cavities are best treated on an individual basis. Shallow caves can be cleaned out or collapsed and then backfilled (as in Grosch et al., 1987), and deeper ones below the level of construction influence may be left untreated. Smaller cavities at intermediate depths can be tolerated by reinforced strip footings, or spread foundations with reinforced rafts, designed to span any potential collapse.

Larger cavities at critical depth are generally best treated by being filled with concrete; if they are open to the surface, a reinforced concrete cap may also be warranted (as in Mangan, 1985) with designs following those for sealing mine shafts (see Chapter 6). Caves may swallow enormous quantities of grout, with major potential losses into void extensions which do not need to be filled. Entering a cave and selectively sealing outlets has proved worthwhile before grout filling on some Yugoslavian construction sites (Nonveiller, 1982). In other cases, forming columns with high viscosity grout can prove economical, and extensive cave development may warrant treatment with perimeter and infill grouting as used in old mine workings (see Chapter 6).

Piles can fail by punching through an underlying cave roof even in misleadingly strong limestone, and micropiles with low individual loadings can therefore offer benefits on cavernous rock. Driven piles generally require at least 3 m of good rock beneath them (Raghu, 1987), and this may need to be proved by a probe, which can be conveniently drilled down between the flanges of an I-section steel pile. Heavily loaded cast piles may warrant proof of 4–5 m of sound rock beneath each one. Cavities below and just to the side of a foundation base can also offer a hazard, and major buildings on limestone in the Hershey Valley, Pennsylvania, are founded on caissons whose integrity has been checked by drilling vertical holes beneath each, and also inclined holes splayed out at 15° (Fosse and Humphreville, 1979). In northern Greece, a bridge was satisfactorily founded on cast-in-situ piles placed right through a limestone cave; a reinforcing I-beam was placed to vertically span the cave, and concrete was then poured into a canvas casing which expanded to twice the diameter of the pile and set to form a column through the cave (Sotiropoulos and Cavounidis, 1979). Caves may also be spanned with steel piles placed through them and bearing on the floor, though grouting and filling is generally cheaper.

Many buildings stand today on cavernous ground, with the benefit that most cavernous rocks are reasonably strong, so, when isolated cavities are found and then filled or spanned, they provide stable bearing capacity. But a single cavity which remains unknown may still constitute an engineering hazard totally out of proportion to its size. Thorough site investigation, based on geological awareness, is therefore of prime importance.

3 Sinkholes on limestone

Diagnostic of karst terrains are surface hollows which drain underground, and, other than on a permeable karstic rock or in a desert, would fill up to form lakes. They are known as dolines, closed depressions or sinkholes. Though in England the word sinkhole generally describes a depression with a visible stream sink in it, the American usage of the word covers all such hollows, with or without stream sinks, and this now dominates in the engineering literature.

Sinkholes nearly all form on terrains of limestone or dolomite (which in this context is merely a variety of limestone), or where either of these rocks occurs not far below the surface. They can, however, form over any rock which is soluble or cavernous (see Chapters 2, 4 and 7). Individual sinkholes may be less than one metre, or more than 100 m, in both depth and diameter, may be circular or elongate, and can have profiles which are conical, cylindrical, saucer-shaped or irregular.

Four main types of sinkholes can be recognized (Figure 3.1). The solution sinkhole forms by slow surface erosion and is the karst equivalent of a valley; the hazard which it offers to foundation engineering lies in the fact that cavities of some size must exist directly or obliquely below it, and these can promote subsidences and other types of sinkhole in its floor. The collapse sinkhole is created by bedrock failure (see Chapter 2), and is rare except on a geological time scale. The buried sinkhole is either of the two previous forms filled and obscured by sediment, and, along with the pinnacled rockhead common on limestone, creates difficult and potentially hazardous ground conditions. The subsidence sinkhole is formed by failure of a soil or weak rock into underlying cavernous limestone; it is far the most abundant type, and its rapid development makes it the major engineering hazard.

The distribution of limestone sinkholes is worldwide, with notable concentrations in southeast Asia, the eastern USA and parts of Europe. Their impact on foundation engineering has promoted an extensive literature, mostly concerned with the subsidence sinkholes. Useful publications, including many case histories, have emanated from some recent conferences, particularly two in Florida (Beck, 1984; Beck and Wilson, 1987), and the 1981 meeting in Istanbul (recorded in Bulletins 24 and 25 of the International

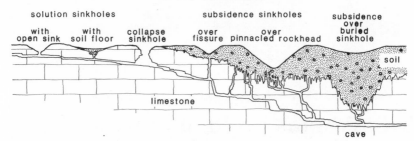

solution sinkholes | subsidence sinkholes | subsidence over buried sinkhole

with open sink | with soil floor | collapse sinkhole | over fissure | over pinnacled rockhead | subsidence over buried sinkhole

soil

limestone

cave

Figure 3.1 Major types of sinkholes and rockhead structures on limestone.

Association of Engineering Geology), while a concise overview is presented by Beck and Sinclair (1986).

Subsidence sinkholes

Subsidence or collapse of a soil overburden into the fissures and caves of an underlying limestone creates subsidence sinkholes without involving failure of the rock. The cover material may be alluvium, boulder clay, clay or almost any other soil, or in some cases may be consolidated rock. The resultant ground hollows, whether formed rapidly or slowly, are best known as subsidence sinkholes, as other terms applied to them (including alluvial, ravelling, subjacent or suffusional sinkholes, and also shakeholes) are too restrictive or localized in use.

The depth of a subsidence sinkhole is limited to the soil depth, and most stabilize to a roughly conical shape with the apex close to the limestone fissure below. The width relates to the soil slope stability; it may be over five times the depth in loose sands in Florida, or less than three times the depth in cohesive boulder clay in England. Many sinkholes in thin soils are less than a metre across, but others over 100 m wide have formed by recent collapse in Alabama (LaMoreaux and Newton, 1986) and Florida (Jammal, 1986).

Some limestone terrains have a density of old sinkholes high enough to influence land use. In parts of Georgia, sinkholes occupy 20% of the ground area (Brook and Allison, 1986), and this figure is matched elsewhere. Additionally, Newton (1987) estimates 6500 new sinkholes have formed since 1950 in the USA, mostly in the southeast, together with many more unrecorded, and structural repairs have run to hundreds of millions of dollars. The frequency of sinkhole collapses in the developing parts of north Florida has had such an impact that insurance companies initiated funding for the Sinkhole Research Institute at Orlando (Beck and Sinclair, 1986).

The mechanism responsible for subsidence sinkholes is the downwashing of sediment by rainwater draining towards a bedrock fissure, though there are variations on this theme. The sinkhole develops by periodic collapses, flow or slumping of the soil, and rapid surface failure is only the conclusion of much

Figure 3.2 A freshly developed subsidence sinkhole about 5 m across, with its area increasing as the soil continues to slump into a fissure in the buried limestone. This has formed naturally in open grassland on the limestone plateau of Guizhou, in China, and has not been induced by man's activities.

Figure 3.3 Debris cone beneath a roof fissure in a limestone cave about a metre high, in the Mendip Hills of southern England. The cave was entered through another soil collapse in the floor of a pipeline trench. There was no surface feature visible at the time above the debris cone, but the soil removal indicates that a subsidence sinkhole may develop in the future (Photo: P.L. Smart).

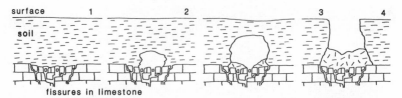

Figure 3.4 Progressive development of a cavity in a cohesive clay soil leading to the catastrophic dropout failure of a subsidence sinkhole. The soil is washed down into the limestone fissures, but there is no failure or collapse of the limestone itself.

slower underground processes. Soil cohesion determines the size any soil cavities can mature to, and hence the size and speed of subsequent failure.

In a cohesive clay soil, sediment downwashing creates cavities which progressively migrate upwards and eventually promote a sudden dropout sinkhole collapse (Figure 3.4). This is very similar to crown hole development over old mines (see Chapter 6). The scale of failure was demonstrated by a 1985 collapse in 21 m of sandy clay above a limestone in Florida (Gordon, 1987); a house fell into a hole 15 m deep and wide within just a few minutes, and after a few hours the sinkhole walls had slumped so that the hole was 18 m across and the house had disappeared. Soil cavities a few metres across are often revealed by foundation work. A monitored excavation near Guilin, in China, showed that most soil cavities were close to the fissured rockhead and within the zone of water-table fluctuation (Yuan, 1983); furthermore, the majority were in silty Fe–Mn clays with very few in stiff clay. Progressive flaking of the roof, in response to water seepage, was observed in a soil cave in South Africa (Jennings, 1966). With continuous washing of the fallen soil from the cave floor down into the limestone, the absence of cavity closure by debris bulking means that subsequent surface failure is inevitable. The speed of the ultimate dropout relies on the cohesion of the cover, and on the small scale often evolves to its most destructive under a tarmac surface.

In a less cohesive sandy soil, with a lack of piping and cavity formation, a sinkhole develops by slow subsidence. Downwashing is still the major process, but continuous soil slumping ensures that a large dropout cannot occur. Fluidization of a low-density soil may also be contributory; Howell and Jenkins (1985) found this common over salt where solution fissures are rapidly enlarged (see Chapter 7), but it can play only a secondary role in modern subsidence over stable limestone fissures.

It is significant that, in both styles of subsidence sinkhole, the ground failure is totally unrelated to cavity enlargement in the bedrock. Limestone solution rates are extremely slow, and surface failure is only a function of soil movement. Bedrock caves do exist beneath some sinkholes, but their role is merely to swallow the debris (Figure 3.5), and sinkholes can develop above only narrow rock fissures, given the time to transmit the soil through them previous to a much faster surface collapse.

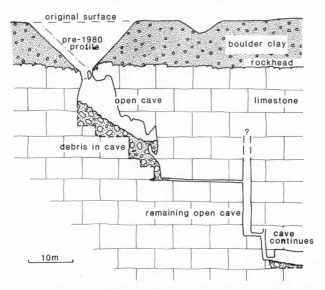

Figure 3.5 A subsidence sinkhole with boulder clay debris washed and fallen into the cave below. A simplified profile of the collapse which occurred in 1980 at Marble Pot, in the Yorkshire Pennines, where the cave had been previously mapped as it was accessible through another sinkhole; the cave is now totally blocked except that water still drains through. The shaft within the cave, whose top has not been seen, could underlie a future surface subsidence, but only if open fissures reach up to the rockhead.

Subsidence sinkholes are generally less frequent where soil depth increases, but variations in drainage and permeabilities prohibit delineation of widely applicable limits. Of the recent sinkholes in China's Shuicheng basin, 60% have formed in soils less than 5 m thick and only 15% in soil thicker than 10 m (Yuan, 1987), while in the Yunxi area 85% of the numerous sinkholes formed in soils less than 4 m deep (Song, 1986). In contrast, the majority of sinkholes in Missouri, USA, are in soils of 5–20 m thickness (Aley *et al.*, 1972), and in Florida, though they are more abundant in the thinner soils, sinkholes are still a significant hazard in soils around 50 m thick. In South Africa, soil mantles over 100 m thick have failed as sinkholes over the limestone of the Rand (Brink, 1979), but these were in the face of induced drainage on an unusual scale for the benefit of the underlying gold mines.

Limestone solution beneath rock sequences, known as interstratal karst, may promote surface failure from even greater depths. In Wales, Thomas (1974) recorded over 400 subsidence sinkholes in strong sandstone overlying limestone at depths up to 160 m, and collapse structures in Missouri were promoted by limestone solution at depths around 220 m (Gentile, 1984). However, these sinkholes are all old features, and there is negligible threat to engineering works over rock sequences as thick as this.

Some sinkholes occur in lines which trace major fractures in the buried limestone, but any widespread pattern is seldom recognizable. A more

Figure 3.6 The 1980 collapse in the subsidence sinkhole over Marble Pot in the limestone karst of Yorkshire. The wide, open fissure in the bedrock can be seen in the shadows beneath the slopes of slumped boulder clay.

significant control of sinkhole distribution is commonly a buried outcrop boundary; soil water draining over and off a shale rockhead onto a buried limestone promotes sinkhole formation in the overlying soil even from a depth of 30 m, as demonstrated by the 1986 Macungie sinkhole collapse in Pennsylvania (Dougherty and Perlow, 1987). In topographical terms, sinkholes are found to occur most frequently in areas of low relief and dry valley floors, but rarely at base level where there is a high (and undisturbed) water table.

Numerous studies of sinkhole distribution indicate minimal opportunity for predicting location of future subsidences or collapses. More success with prediction, which is useful in an engineering context, is achieved through recognizing the immediate causes of sinkhole failure. Only a small number are of purely natural origin, and these are commonly related to rainfall patterns. The vast majority of modern sinkhole collapses are induced by man, through some form of engineering or construction activity (Newton, 1986),

and of these the most influential factor is a falling water table due to overpumping of the limestone aquifer.

Sinkholes induced by water-table decline

Abstraction of groundwater and the consequent water-table decline has
induced sinkholes which account for some of the world's most dramatic and
destructive ground subsidence events. A clear correlation, in both space and
time, between water-table decline and subsidence sinkhole failures has
been demonstrated in many limestone areas where groundwater has been
abstracted for supply or to drain bedrock around mines. One guideline that
repeatedly emerges from case studies is that sinkholes form most often where
and when the water table first declines past the rockhead.

An early classic case of sinkhole development was in the Hershey Valley of
Pennsylvania, where a limestone floor is overlain by 20–30 m of soil. In 1949, a
limestone mine, in order to prevent its own flooding, increased pumping and
created a cone of depression where the water table fell at least 10 m over an
area of 600 ha. Within three months, 100 subsidence sinkholes formed in the
soils of the same area, and their development threatened the nearby chocolate
factory. Subsequent events—the chocolate company recharging the aquifer,
the extensive litigation, and the grouting of the limestone around the mine
(Fosse, 1969)—ended only when the mine was bought, closed and flooded to
restore the water table and remove the sinkhole hazard.

A modern equivalent of Hershey is the Dry Valley area of Alabama, where a
limestone mine has lowered the water table over 100 m in an area with soils
generally 5 m deep over a pinnacled rockhead (LaMoreaux and Newton, 1986).
Above the cone of depression, several thousand sinkholes, mostly 1–30 m
across, have formed in the period 1967–84, and one road across the valley
has needed repairs over 30 times within five years. Many other cases of
subsidence sinkholes induced by water-table decline are documented by
Newton (1984, 1986, 1987). In north Florida, 70% of the sinkhole collapses
occur in April and May at the time of maximum groundwater abstraction
(Beck and Sinclair, 1986), and a Tampa wellfield induced 64 new sinkholes
inside a 1600 m radius within two months of increasing pumping rates
(Sinclair, 1982). These subsidences are so common in the USA that the
groundwater pumper is now liable for subsidence damage outside his
property; the legal situation concerning pumping-induced sinkholes is
reviewed by Quinlan (1986).

Water-table decline induces sinkhole development mainly through erosion
by the increased downward flow of the soil groundwater. An increase in
drainage velocity also accelerates soil piping, especially where turbulent flow is
created. Clearly, an induced water-table decline, when the soil-water flow is
maintained, can promote more sinkholes than a water-table fall during a
drought with decreased soil drainage. Loss of buoyancy support by the

B

Figure 3.7 Cautionary signs on the road across Dry Valley in Alabama, where the frequency of sinkhole failure has been greatly increased by a pumped decline of the water table.

groundwater aids soil cavity failure, and vacuum suction may even be created beneath a clay soil (Yuan, 1987). Water-table fluctuations may cause further erosion of a clay by cyclic wetting and drying. Sinkholes are then most effectively induced after a water-table decline past the rockhead initiates all these processes at the bottom of the soil profile. Conversely, the only really effective way to control a phase of sinkhole activity induced by overpumping is to let the water table recover, though this is not easily achieved where there is regional decline through groundwater resource exploitation, and may necessitate a grout curtain around an active mine.

One of Florida's largest recent collapses was the Winter Park event in 1981 (Jammal, 1984), where nearly 150 000 m³ of soil dropped away to create a sinkhole 105 m across and 30 m deep. It formed in soil 45 m deep over a limestone whose water table had declined 6 m in the previous 50 years. The water table had not fallen to the rockhead, but its decline had increased the hydraulic head across the aquiclude in the lower soil sequence between the aquifers above and below it (Figure 3.8), and it was the failure of the aquiclude clay which prompted the collapse. The sinkhole pond first stood at the level of the limestone water table, but, after some fluctuations, it stabilized at the sand water-table level when the underlying pipe blocked itself with fallen soil.

In China, there are eight sites each with over 1000 sinkholes induced by pumping since 1975 (Song, 1986). Seven are in mining areas of Hunan and Guangdong. The other is in the Shuicheng basin of Guizhou where water has been abstracted for industrial supply (Waltham, 1986). Soils in the basin are mostly less than 10 m thick, and pumping from 17 wells in the limestone has

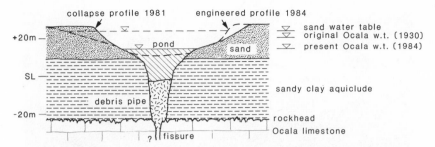

Figure 3.8 Profile through the Winter Park sinkhole which collapsed in Florida in 1981 (after Jammal, 1984).

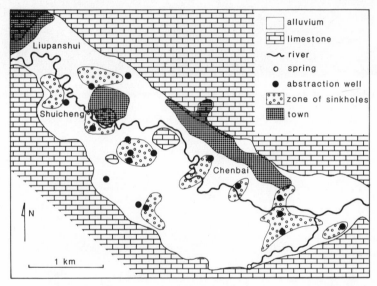

Figure 3.9 Map of the Shuicheng basin in Guizhou, China, where hundreds of sinkholes have developed in the thin alluvial cover on bedrock limestone. Most of the sinkholes are in well-defined zones which show a clear spatial relationship to recently exploited groundwater abstraction wells (after Waltham and Smart, 1988, and Kang, 1984).

locally lowered the water table by 20 m. Sinkholes have damaged many houses and buildings, and there is a clear spatial relationship between their distribution and the pumped wells (Figure 3.9). The uneven shape of the sinkhole zones may reflect fissures or the main groundwater routes in the limestone. Significantly, most sinkholes formed within a few days after the adjacent well was pumped for the first time.

Experience in China has allowed Song (1986) to recognize three concentric zones around pumped wells, with the inner, most prone to sinkholes, defined by the water table falling past the rockhead. Over a cone of depression around a Guangdong mine the sinkholes occurred mainly above faults, along river

Figure 3.10 One of the many sinkholes induced by groundwater pumping from the fissured limestone beneath the alluvial floor of the Shuicheng basin in China. Pinnacled limestone is revealed only a metre below the original surface, and also forms the steep conical hills in the background.

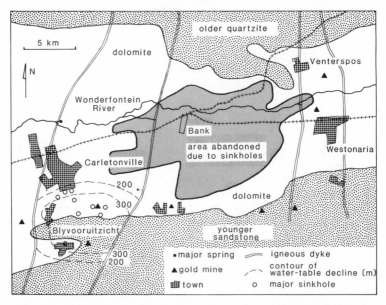

Figure 3.11 Map of the Far West Rand of South Africa, which has been seriously affected by sinkholes since the cavernous dolomites were dewatered to facilitate operations in the underlying gold mines. The igneous dykes divide the dolomite into hydrological compartments. Dewatering of the Bank compartment caused so many sinkholes that much of the ground on the dolomite outcrops, including the town of Bank, was abandoned. Contours of water-table decline, and associated major sinkholes, are only shown in the Oberholzer compartment which contains the town of Carletonville.

banks or in wet depressions (Dai and Zhu, 1986), all where drainage through the soil is maximized; but over a similar mine in Hunan a 50-m soil cover prevented sinkhole formation, though the pumping induced failures in soils 2–3 m thick over the same limestone 4 km away (Song, 1986). Clearly sinkhole distribution is influenced by both water-table decline and local details of the geology.

Even more dramatic than the sinkholes of the USA and China are those on the dolomites of the Rand in South Africa. Since 1957, these have killed 38 people, destroyed over 100 buildings, prompted evacuation of a township (see Figure 3.11), derailed a train, and created damage worth many millions of pounds; the events are well documented by Bezuidenhout and Enslin (1970) and Brink (1979, 1984). Hundreds of sinkholes have developed, including dropouts 30 m deep and slow failures 150 m across, all promoted by pumped drainage of the thick karst dolomites for the benefit of the world's richest gold mines which lie immediately beneath. Some of the soils over the dolomite rockhead are over 100 m deep, but with a pumped water-table decline locally greater than 300 m, subsidence sinkholes have formed on a proportionate scale which is hardly matched elsewhere.

Sinkholes induced by environmental disturbance

A large proportion of sinkhole failures are induced by engineering activities and are of major significance because they directly affect the site being developed, either immediately or some years later. A declined water table may have promoted sinkholes directly, or left sections of ground in a critical state awaiting the disturbance which triggers their failure; even without a water-table decline, the same disturbance may prompt failure, but statistically less often.

The engineering activities which can trigger sinkhole collapses are many and disparate, and are best summarized as a list of hazardous processes:

(i) Excavation of part of a soil cover may thin the roof of a soil cavity and so accelerate a failure; and removal of a clay soil may permit drainage through previously sealed sands. Some Missouri railroads stand on banks made from soil excavated adjacent to them, and the marginal hollows frequently develop sinkholes (Aley et al., 1972).

(ii) Removal of vegetation permits increased infiltration and also deprives the soil of its root mat. In Alabama, sinkholes are more common in the parts of Dry Valley where timber has been cut (LaMoreaux, 1984), and failures occurred on a Birmingham construction site when only foundation trenches broke areas stripped of topsoil (Newton and Hyde, 1971).

(iii) Excavation by blasting may disturb soils over a considerable area. The village of Liangwu, in southern China, was abandoned when nearby blasting triggered 40 sinkholes, and another 100 followed soon after in an area 1800 m long (Yuan, 1983, 1987).

Figure 3.12 Small subsidence sinkholes in a soil over limestone causing failure of the surface of a shopping centre car park in Bowling Green, Kentucky. The patching of the surface indicates the extent of earlier collapses, and the drain near the fence is clearly inadequate for its task.

(iv) An unsealed borehole may allow a perched soil aquifer to drain into a bedrock cave. A drill rig was lost into a self-induced sinkhole 20 m deep in Florida, in 1974, and within two days another dozen smaller sinkholes had formed up to 100 m away.

(v) Imposed load, either structural or from site plant, may precipitate a failure, which would probably have occurred later anyway.

(vi) The major influence from construction or any form of urbanization is the diversion of natural drainage. Runoff from roads or buildings usually concentrates drainage at certain points, where sinkholes may develop as soil erosion is enhanced and clay cohesion is reduced.

(vii) Disposal of runoff into soakaway drains, or dry wells, in a soil over limestone is a prime cause of failure, often beneath the building or road being drained. Drainage wells cased into the limestone should perform safely, but, if poorly installed, leakage may cause adjacent or nearby failures (Crawford, 1986).

(viii) Unlined drainage ditches also create point discharge into the soil. In Pennsylvania, 7 km of highway induced 184 sinkholes within 12 years, all along its associated drainage channels (Myers and Perlow, 1984).

(ix) Sinkhole collapse may be induced by water flowing from a pipeline fracture, which itself could have been caused by soil movement as a slow subsidence sinkhole developed. This can be destructive to urban structures served by the pipeline (as in Canace and Dalton, 1984).

(x) Irrigation water added to a soil can often induce sinkholes. The garden hose was often left running overnight at the site of the Blyvooruitzicht

Figure 3.13 A subsidence sinkhole developing adjacent to a well-head housing in the Shuicheng basin of China. The soil subsidence is induced by both the pumped cone of depression in the water table and also pipeline leakage draining back into the ground. This recurring hazard had been recognized elsewhere in the basin, so the building is founded on piles bearing on to stable rockhead.

collapse which killed five people in South Africa (Brink, 1984). The disastrous practice of leaving a flowing standpipe to supply water for the local peasant farmers has promoted a number of the sinkholes in China's Shuicheng basin (Waltham, 1986).

(xi) Reservoir impoundments over limestone initiate sinkholes most effectively; there the problems are more of leakage than subsidence, as testified by empty reservoirs in Missouri (Aley *et al.*, 1972), besides others in China, Indonesia and elsewhere.

Soil drainage increases naturally in response to heavy rainfall, and a correlation between sinkhole failures and storm events has been recognized in South Africa (Brink, 1979), Puerto Rico (Wegrzyn *et al.*, 1984), Belgium (Delattre, 1985) and the sandy soil areas of Florida (Beck and Sinclair, 1986). The few recent sinkhole collapses in Britain's limestone were all associated with unusually heavy rainfall. In consequence of a storm event, a rapidly rising water table may also induce sinkhole failure by the escape of trapped air from limestone voids beneath a clay soil, especially under reservoirs (Chen, 1986).

Conversely, a clay soil may develop desiccation cracks during a drought, and thus be ready to fail over any void with the first renewal of rainwater flow. The Liangwu collapses in China were precipitated by blasting, but did occur at the end of a long drought (Yuan, 1983), and Newton (1986) records drought-related sinkholes in Alabama, Florida and Tennessee. The similar consequences of both wet and dry soil conditions may account for the lack of correlation between weather and sinkhole failures in the clay soils of China's Shuicheng basin, and Beggs and Ruth (1984) similarly find no rainfall influence on highway-related sinkholes in parts of Florida. Overall, it would appear that water-table decline and environmental disturbance are the primary factors promoting sinkhole collapse, and rainfall patterns play only a secondary role in that they may influence the timing of events.

Prediction and zoning of sinkhole hazards

Sinkhole subsidence events can rarely be predicted except in the broadest of terms. Even in the well-documented limestone areas of north Florida, the impossibility of prediction makes sinkhole insurance the best defence, and this is now mandatory on residential property (Beck and Sinclair, 1986). Prediction research is restrained because the event frequency, as a hazard to any single site, is extremely low; occurrence rates calculated for part of Florida rise to only one sinkhole collapse per square kilometre per 10 years (Upchurch and Littlefield, 1987), and this is probably close to a maximum rate except for areas of rapid water-table decline.

It is, however, possible to recognize zones where the sinkhole risk is increased. Defined by the major controlling factors, the most hazardous zone is a valley floor, with limestone bedrock beneath a soil cover 2–20 m thick (though local data can modify these figures), where the water table is declining past the rockhead. This zone will not enclose all sinkhole events, but it is probably the only widely applicable delimitation with a hazard potential worthy of planning consideration. The hazardous soil thickness limits vary between karsts, and the best local guideline is the distribution of existing subsidence sinkholes, as new collapses are rare in previously undisturbed areas. Soil profiles may add another parameter to the zone definition; in Kuala Lumpur, Tan (1987) recognizes a pattern of a stiff clay overlying loose piped sand above a pinnacled rockhead, which should be locally zoned out with respect to future construction.

The greatest applications of sinkhole hazard zoning has been on the Rand dolomites over the South African goldfields. Along the floor of the Wonderfontein Valley (Figure 3.11), hazard zones have been defined as the areas of deep pinnacled rockhead together with their marginal steep rockhead slopes (Kleywegt and Enslin, 1973) which have been mapped by their negative gravity anomalies (see below). These zones encompass the widespread slow subsidences, typically developed over buried sinkholes (see below) and also the highest risk of new collapse sinkholes. The subsidences mostly occur when mine dewatering lowers the water table within the zone, and this occurs in discrete blocks along the dolomite outcrop as the aquifer is divided into hydrological compartments by watertight igneous dykes. The zoning is not relevant to individual building threat, especially as the collapses are still randomly distributed within the zone, but is of major cost benefit in siting new

Figure 3.14 A house in Bank, South Africa, destroyed by severe subsidence and ground fissuring as a result of dewatering the cavernous dolomite with its large buried sinkholes beneath a thick soil cover (Photo: W. Gamble).

urban development (Partridge *et al.*, 1981; Brackley *et al.*, 1986; Venter and Gregory, 1987). New townships are excluded by law from the hazard zones, and a recognized zone around Bank has been evacuated and abandoned (Figure 3.11).

Major rockhead fissures may locate future soil collapses, and conspicuous lines of sinkholes are best avoided by new construction. However, the full pattern of bedrock fractures is obscured by the soil cover; statistical analysis may identify broad zones of increased sinkhole susceptibility (Brook and Allison, 1986), but any more accurate prediction of future collapse locations, based on sinkhole patterns, is usually unrealistic.

Modern airborne photography, especially using multispectral scanners through the visible and thermal infrared ranges with digital enhancement of selected bands, reveals vegetation stress and soil drainage patterns. Over a limestone, circular or annular spot anomalies, due to wet soil in a developing slow subsidence or dry soil drained into a fissure, can be interpreted as potential sinkhole sites, and linear anomalies may trace bedrock fractures. Some of the Florida karst has revealed a correlation between sinkhole activity and airborne images (Coker *et al.*, 1969). However, this technique may involve initial costs which are a barrier to use on a small site, though it can contribute to regional development planning.

An impending sinkhole collapse may offer warning signs. The most important are circular cracks in either soil or tarmac, slow localized subsidence perhaps with building distortion, pool formation in new hollows or loss of drainage, vegetation stress, and muddy water in nearby wells. Newton (1987) extends the list to 14 factors. Once an unstable site is recognized, a telescopic benchmark or extensometer (Marsland and Quarterman, 1974) can be installed in a borehole to monitor elevation changes through the soil profile. Soil contraction implies future consolidation and slow subsidence, while dilation is the precursor to a dropout collapse (Jennings, 1966). Designed for South Africa's large sinkholes, these are of limited practical use at most other sites. Alternatively a geophone can monitor progressive soil failure through an increase in micro-seismic rock noise (Belesky *et al.*, 1987). Even with these refined techniques, the timing of a failure is not easily predicted, except that it will probably be precipitated by some rainfall event.

Precautionary engineering in sinkhole areas

There is no doubt that construction activity increases the rate of sinkhole formation in a soil-covered karst by imposing change in various environmental parameters (listed above). Good practice attempts to minimize these impacts, and pays particular attention to drainage disturbance, especially in areas of water-table decline. Positive action may be appropriate on some difficult sites. This can include surcharging the soil to precipitate small failures before construction, grouting to stabilize the soil cover (Clark, 1961; Belesky *et al.*, 1987; Henry, 1987), grouting the bedrock fissures (Venter and Gregory, 1987), or the use of driven piles instead of caissons, where a water table pumped down while building the latter may induce subsidence sinkholes adjacent to the site (Foose and Humphreville, 1979). In China, holes have been bored through impermeable clay soils to let air into a cavernous limestone during a water-table decline and so eliminate suction effects (Dai and Zhu, 1986).

The prime precaution is to restrict runoff drainage into the soil, especially where the area under pavement or roof is increased; in South Africa it is sometimes considered worthwhile to pave the areas between buildings (Partridge *et al.*, 1981). All runoff must be carried away from structures. Interception ditches around any hollows or old sinkholes can prevent renewed subsidence, and closed basins with internal drainage should be avoided, landscaped out, or only built over when efficient artifical drainage is installed. These basins may only be recognized by careful surveying, as their depth may be less than the contour interval on available maps.

Pipeline trenches should be backfilled with clay compacted till it is less permeable than the adjacent soil, so that they do not act as french drains, in which case they can gather drainage, induce sinkholes and self-destruct. Flexible pipes are best employed, and in areas of major sinkhole hazard pipelines can be left above ground (Partridge *et al.*, 1981).

Drainage ditches require impermeable linings, as attested by the great number of sinkhole failures due to leakage from untreated ditchlines along Tennessee highways, compared with the rare failures under paved ditches

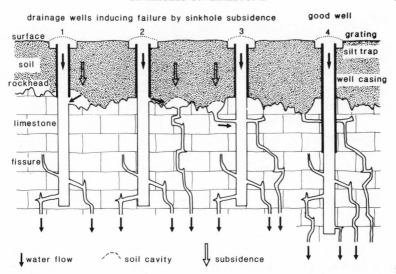

Figure 3.15 Drainage wells through a soil cover into a buried limestone, which may cause sinkhole failures due to poor construction. Wells 1 and 2 have casings which are not sealed on to the rockhead, and can induce soil cavitation by water draining in or out of the wells. Well 3 is sealed on to an almost detached block at rockhead and water flushing in or out of the well can disturb the nearby soil. Well 4 is cased through the weathered and fissured zone of limestone just below the rockhead, and may therefore operate without disturbing any soil (after Crawford, 1984).

(Moore, 1987). Similarly, unlined stormwater retention basins effectively create their own sinkholes to drain into a limestone below. Ideally, dry wells should not be used for runoff disposal in a sinkhole-prone karst terrain. A Pennsylvania industrial unit, sited in a shallow basin in soil-covered limestone, had half its area under roof or pavement and utilized dry wells for storm drainage; soon after completion, heavy rain induced over 20 sinkholes and prompted installation of an elaborate new drainage system, adding 40% to the original site development costs (Knight, 1971). Where the low and disorganized relief of a sinkhole plain leaves no alternative, dry wells can be safely employed if cased down to and into bedrock (Figure 3.15). Experience in Bowling Green, Kentucky, has proved that unless a drainage well is sealed with concrete into a bedrock fissure, and preferably cased to 3–5 m below rockhead, it will induce subsidence and failure as it gathers soil drainage (Crawford, 1986).

Old subsidence sinkholes can be left completely untouched, or should be cleaned out and backfilled with granular material to allow drainage through them and not through the soil around them. Any signs of recent subsidence will demand more elaborate repair (see below). A significant subsidence hazard rises from old sinkholes poorly filled, graded and forgotten; these may subsequently fail, as at the Macungie sinkhole in Pennsylvania (Dougherty and Perlow, 1987) unless found by recourse to old topographic maps, air photographs or soil augering, and renovated prior to construction.

Sinkhole repair

Uncontrolled backfilling of a newly collapsed subsidence sinkhole invariably leaves the way open for reactivation in the future. Some sinkholes in Pennsylvania are refilled almost every year (White et al., 1984), and Newton (1987) cites various cases of renewed subsidence after longer time intervals. On the other hand, low-rise housing units in St Louis, USA, have performed to design over small old sinkholes which were just filled with graded material, though efficient storm drainage was also installed (Reitz and Eskridge, 1977).

Larger sinkhole collapses require a systematic repair which both seals the bedrock fissure to prevent subsequent piping and creates a stable soil profile. Figure 3.16 shows a generalized repair scheme, but there is scope for variation within its components. If rockhead can be exposed at the apex of the collapse, the critical fissure can be sealed with concrete, though this is often not possible, and there is the danger of merely diverting soil drainage to the next fissure. In Florida's giant Winter Park sinkhole, soil was pushed in to reinforce a seal already developing (Jammar, 1984) and its integrity could then be monitored by the subsequent lake level (see Figure 3.8). Normally the cleaned sinkhole is filled to a thickness of 2–5 m with chunk rock coarse enough not to run into fissures, on which a reinforced concrete slab can be formed (as in the Macungie sinkhole, illustrated by Dougherty and Perlow, 1987). Beneath just a shallow soil, a small fissure can be covered with a simple slab founded on bedrock, as if it were a mineshaft (see Chapter 6).

With its throat capped, a sinkhole can be filled to grade. In areas of shallow water table, hydrostatic pressure across the concrete slab can be eliminated by a pipe set in the slab and feeding to a basal filter bed in the fill (Reitz and Eskridge, 1977). A reinforcing layer of geotextile just below the surface can improve safety by ensuring that any subsequent soil movement first develops with only a slow surface subsidence (Bonaparte and Berg, 1987); alternatively, geotextiles can line a small sinkhole and substitute for the concrete slab at its apex (Sowers, 1984). If land use permits, a gentle dome of low-permeability soil over the sinkhole fill reduces subsequent drainage into it. The future behaviour

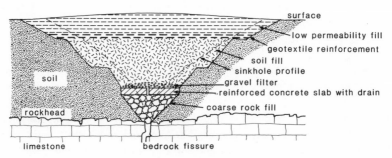

Figure 3.16 Generalized scheme for the repair of a subsidence sinkhole (partly after Bonaparte and Berg, 1987).

of a large active sinkhole is so difficult to predict that repair designs should be conservative, and costs are therefore high. The Macungie collapse, beneath a Pennsylvania road, cost $450 000 to reinstate (Dougherty and Perlow, 1987), and the Winter Park sinkhole cost $85 000 merely to stabilize as a lake (Jammal, 1984), but most sinkholes are smaller than these.

Buried sinkholes

Rockhead relief creates an engineering hazard through the potential for differential settlement where the foundations of a single structure bear on contrasting materials across the site. The roof of a supermarket at Akron, Ohio, failed in 1969 as a direct consequence of one column settling 200 mm more than its neighbours; this was because it was founded on a soft clay within a buried sinkhole which had not been found in the site investigation of the depth to the limestone rockhead.

The selective solutional erosion of limestone creates localized rockhead relief on a scale seldom matched in other rocks. This may form only deep soil-filled fissures, of the type exposed at some dam sites (Soderberg, 1979), or the fissuring may be dense enough to warrant the description of pinnacled rockhead (Figure 3.1). Alternatively, the fissures may widen or coalesce to create buried sinkholes. All these phenomena develop either by pre-burial subaerial erosion or subsequent subsoil erosion, and occur on a larger scale in tropical environments with enhanced solution processes.

Figure 3.17 A pinnacled limestone rockhead exposed in a road excavation in Guizhou, China. The soil cover has been dug out by hand, preparatory to breaking down the rock pinnacles, most of which had their tops within a metre of the original surface.

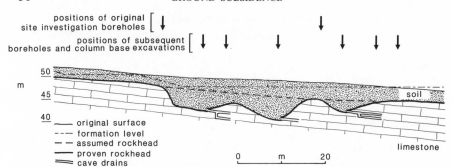

Figure 3.18 Profile through the buried sinkholes revealed in the sloping limestone rockhead beneath a construction site at Chepstow, in South Wales (derived from Statham and Baker, 1986).

Soil depths ranging from 2 to 15 m over a limestone in New Jersey (Canace and Dalton, 1984) typify the relief of a pinnacled rockhead; but pairs of boreholes 5 m apart on a construction site in Kuala Lumpur, Malaysia, found the rockhead changing from 5 m to 80 m deep, and demonstrate the scale which can be achieved. Rockhead pinnacles may have been notched or undercut by past erosion stages (Tan and Batchelor, 1981), and may be detached from the bedrock so that their only lateral support is the soil. Eventually they may collapse, and the intervening fissures remain as a chaos of soil, slipped blocks (often described as boulders or floaters) and soil cavities, creating difficult foundation conditions. At a Chicago power station, the safe bearing pressure was taken as 4 MPa for the solid bedrock limestone, but only 1.5 MPa for the zones of pinnacles, soil and rubble (Swiger and Estes, 1959).

An individual buried sinkhole presents a subsidence hazard where it has a weak soil infill and has no surface expression. Even small features, such as those less than 20 m across and 5 m deep (Figure 3.18) revealed on a supermarket site in South Wales (Statham and Baker, 1986) can threaten foundation performance if not recognized before or during construction. Furthermore, the scale of rockhead relief is unpredictable; construction of the Newton Abbot bypass, in southwestern England, was expected to reveal soil 2 m deep over limestone, but some bridge re-design was then necessitated by a series of buried sinkholes over 20 m deep with rockhead slopes of 1:1 (Low and Bramwych, 1971). In South Africa, an area 250 m across subsided to 10 m in the centre, during a period of three years, above a buried sinkhole over 50 m deep; the movement was prompted by the water-table decline over the Rand gold mines (Bezoudinhout and Enslin, 1970). This was just one of many slow sinkhole subsidences over the dewatered dolomites of the Rand, the worst developing over rockhead depressions filled with very weak manganese wad soils. In this style of subsidence, the most destructive differential movement occurs over the steep rockhead slopes of the buried sinkhole margins.

Site investigation over buried sinkholes and pinnacles

A borehole investigation of a pinnacled rockhead proves difficult because there can be endless and unpredictable variety in the pinnacle profiles. Close borehole spacing is required and, especially in tropical karsts, increased borehole numbers merely reveal even more pinnacled complexity. A grid of holes can be meaningless, and at least one borehole must be placed at the site of every column foundation. Bores must prove bedrock, as opposed to a detached pinnacle or soil-supported boulder, by penetrating the rock, usually to 4–8 m, to clear the maximum boulder size generally found in the area. Inclined boreholes more easily find vertical fissures or dangerously thin pinnacles, and, if soil cavities are known to exist, air track drilling is advisable, as water flush may induce a subsidence sinkhole beneath the rig.

Where pinnacle tops are close to the surface, soil stripping or trenching may reveal their distribution better and more cheaply than a large number of boreholes. This was done at the Zeerust site in South Africa (Figure 3.21) where 48 boreholes had revealed no pattern in rockhead depths ranging from 0.3 m to over 20 m (Brink, 1979). Where a bulk site excavation is planned, the deeper boreholes are better sunk after than before, to take account of any bedrock profiles revealed (Swiger and Estes, 1959). Boreholes have a statistically higher chance of finding large buried sinkholes, but can still miss significant features. The buried sinkholes shown in Figure 3.18 were missed on the initial investigation, and were found only when column bases were excavated; fortunately, construction was not delayed while piers were deepened, and a more extensive initial exploration could hardly have been justified (Statham and Baker, 1986).

Clearly, a site investigation on soil-covered limestone must adapt to local conditions, and be constantly modified as data are gathered. High drilling costs justify consideration of other techniques. Airborne imagery, especially covering the infrared spectrum (Edmonds et al., 1987), may reveal spot anomalies over buried sinkholes, and conventional air-photograph coverage at least is readily available for many areas, though its value is limited on urban sites.

Geophysical profiling of pinnacled rockhead suffers from many of the limitations outlined in Chapter 2 with reference to the locations of cavities. Conventional resistivity traversing was used by Early and Dyer (1965) to find bedrock fissures beneath a soil cover on a site in the Derbyshire Pennines. They interpreted low resistivity anomalies as clay-filled fissure zones, but correlation with reality revealed by subsequent excavation (Figure 3.19) showed they could do no more than indicate sound rock within their positive anomalies. Their survey took seven days, but would be faster and cheaper today with modern electromagnetic systems. Ballard et al.(1983) also found the technique useful on a dam site investigation in Guatemala.

Large buried sinkholes create negative gravity anomalies due to the low density of their soil infills. The subsidence-prone karst of the South African

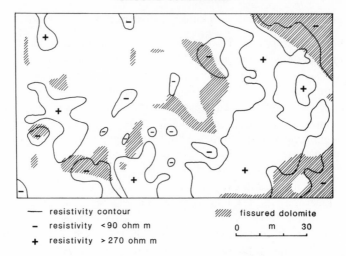

— resistivity contour

− resistivity < 90 ohm m

+ resistivity > 270 ohm m

▨ fissured dolomite

0   m   30

Figure 3.19 Results of a geo-electric survey, correlated with the distribution of fissures subsequently revealed by excavation, on a site on soil-covered karst dolomite near Matlock, Derbyshire. The clay-filled fissures within the bedrock could be expected to create negative anomalies in contrast to the unfissured high-resistivity dolomite, but the results show only a qualified success for the geophysical survey (after Early and Dyer, 1964).

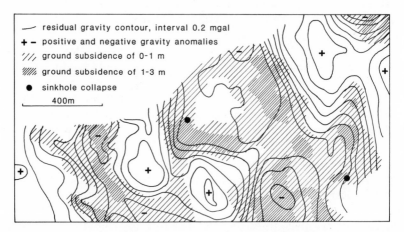

— residual gravity contour, interval 0.2 mgal

+ − positive and negative gravity anomalies

/// ground subsidence of 0-1 m

▨ ground subsidence of 1-3 m

● sinkhole collapse

400m

Figure 3.20 Correlation of a gravity survey with subsequent subsidence across part of the town of Bank in the Rand goldfield of South Africa. Positive gravity anomalies over shallow dolomite bedrock contrast with negative anomalies over large and deep buried sinkholes; subsidence is due to compaction of the sinkhole sediment fills when dewatered (after Kleywegt and Enslin, 1973).

goldfields is regionally zoned by gravity surveys (Kleywegt and Enslin, 1973). Negative anomalies greater than 1 mgal identify rockhead depressions over 50 m deep where soil compaction causes subsidence of at least 1 m when the water table declines due to mine dewatering. Figure 3.20 shows the usable correlation between gravity data and subsequent subsidence, but few other areas attain these scales of buried sinkholes and subsidence. Negative anomalies of 90 μgal traced buried sinkholes 3–4 m deep in a chalk rockhead below 3–4 m of soil at a Belgian airport site (Calembert, 1975), but the survey was of questionable practicability.

Of other geophysical methods, seismic refraction surveys of rockhead are generally too imprecise for pinnacle recognition, while positive magnetic anomalies may identify clay-filled pockets in chalk (McDowell, 1975). Ground radar has recognized pinnacled rockhead in Florida (Gilboy, 1987) and subsiding zones within the soil profile in a North Carolina karst (Benson and Yuhr, 1987), but in both cases is limited to depths of 6 m. Overall, the available remote-sensing techniques cannot prove a sound rockhead free of buried sinkholes or pinnacles, but can in some cases reveal anomalies which provide guidelines to a more economical borehole investigation.

Foundations over pinnacles and buried sinkholes

A very uneven limestone rock surface beneath a soil cover demands careful foundation design for any but light structures which can be safely founded within a thick soil. Dynamic compaction can improve a soil in the 3–20 m depth range by collapsing soil cavities (Guyet, 1984), but this does not entirely remove the chance of future piping and subsidence, even in the less permeable compacted soil. Excavation to a shallow pinnacled rockhead and placement of a coarse rock backfill can achieve a firm footing, or a crushed rock mattress can be established in the soil (Wagner and Day, 1986). In either case a concrete floor slab formed on the fill should be reinforced so that it can safely bridge any soil cavity that may reasonably be expected to form beneath. Generally, a soil over 15 m deep removes any direct subsidence hazard, unless there is a subsequent water-table decline, as was the case with the South African gold-mine crusher which failed even though it was founded on a grouted soil mat over a deep rockhead (Brink, 1979).

Where pinnacle tops lie at accessible depths, they may support structural loads once their integrity has been proven by exploratory boreholes. In South Africa, the Zeerust television station was built on ground beams designed to bear on pinnacles exposed in a shallow excavation (Brink, 1979), and some beams were extended to reach stable pinnacles (Figure 3.21). Reinforced ground beams can similarly span some buried sinkholes, and designs which specifically account for the distribution of pinnacles have permitted successful founding of a pipeline (LaMoreaux and Newton, 1986) and part of a power station (Swiger and Estes, 1959), both in the USA. Other parts of the same Chicago power station were supported on caissons taken to depths of 18–40 m, to reach below the pinnacles and found on solid limestone with a higher bearing capacity. A similar caisson foundation was designed for a hospital in the Hershey Valley of Pennsylvania, where careful site investigation allowed selection of an area of shallow rockhead; a mean caisson length of 4 m was

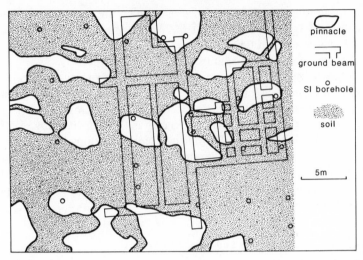

Figure 3.21 Plan view of foundations designed to fit the morphology of dolomite pinnacles beneath the Zeerust television station in South Africa. The pinnacle tops were less than 1 m below the surface, but the soils between the pinnacles reached depths of over 20 m (after Brink, 1979).

predicted, but eventually proved to be 5.5 m, and since then a safety margin of about 30% has been added to foundation costings on comparable limestone sites (Foose and Humphreville, 1979).

Steel piles can easily suffer from deflection, bending, curling or poor seating where driven to a pinnacled rockhead (Sowers, 1984), though they can be driven through any rubble zone or even into a weathered limestone to find good rock. On deeply pinnacled rockhead, such as under Kuala Lumpur, steel piles driven to resistance, at whatever depth they find sound rock, offer the most adaptable foundation, and can achieve capacities around 1000 kN (Bergado and Selvanayagam, 1987). Driven concrete piles are more difficult to place on pinnacles, and more numerous small-diameter piles provide safer load distribution. Alternatively, bored piles can be placed a few metres into bedrock, where fissure densities decrease away from rockhead, though pile lengths may vary considerably where buried sinkholes exist (as in Calembert, 1975). In view of the difficulty of proving the integrity of buried limestone for any piled foundation, reinforced slabs or beams are best designed so that they can survive the failure of any one pile support.

Solutional erosion of limestone presents an infinite variety of ground conditions, and Sowers (1984) demonstrates a wide range of foundation structures each suitable for a specific situation. Sweeping guidelines are inappropriate, and empirical designs, based on local experience and then adapted as both investigation and excavation yield more data, prove the most successful.

4 Subsidence and collapse on chalk

Chalk is a fine-grained, soft, pure, white and porous variety of limestone. It has an extensive outcrop across northern Europe, including nearly 15% of the solid geology of England. Chalks also occur elsewhere in the world, including the Caribbean, the Middle East and Australia, and though some of these are not of Cretaceous age, they are all geologically young. The distinctive feature of chalk is that, like other limestones, it is slowly dissolved in natural water and can therefore contain cavities, and yet it has a much lower mechanical strength, especially when weathered. Chalk forms its own characteristic style of karst landscape, with rolling hills, dry valleys and underground drainage, but with a noticeable lack of bare rock crags (except in undercut sea cliffs).

Over 75% of chalk typically consists of shell structures, known as coccoliths, which are less than 5 μm in diameter. The high porosity of the chalk, ranging from 20 to 50%, is due to the open packing of the coccoliths and also to cavities within them. Chalk's variable physical properties relate to its local history of preconsolidation loading and burial; the weaker beds, notably the Upper Chalk which dominates the outcrop in England, have never had a cover of Tertiary rocks. Joints and fissures in the chalk are irregular, and are widely spaced at depth, but their density increases significantly up through the surface-weathered zones.

Unconfined compressive strength of dry unweathered chalk ranges from 5 to 27 MPa. However, saturated strength is reduced to between 1.7 and 12 MPa, with the most conspicuous loss exhibited by the very porous Upper Chalk; with its fine grain size and high water retention, most chalk is saturated. The high porosity of chalk also accounts for its spectacular susceptibility to frost shattering. Natural weathering vastly increases the fracturing and disintegration of the chalk, and creates a surface layer of structureless 'putty' chalk of very low strength. Ground subsidence over chalk may therefore be related to either the formation of solution cavities or the failure of putty chalk.

Caves, pipes and sinkholes

Chalk does not warrant the general description of 'cavernous', because its high fracture density, high porosity and also low strength enable it to act largely as a diffuse-flow aquifer. However, solution caves do exist in chalk, as at the Water End sinkholes which swallow a major stream draining part of north London; excavation of the main sink has revealed tubular passages a metre across and partly filled with clay, and the stream cascades down and along a series of fissures too narrow to enter. Isolated caves are known elsewhere, and the stronger chalks of France contain accessible stream caves over 2 km long. Shallow solutional voids do create a hazard to bored piles bearing on buried surfaces of unweathered chalk, and so warrant site investigation practices applicable to cavernous limestone. This does not apply to driven piles in weathered chalk.

Based on the concept of concentrated drainage flow and solution, it has been suggested that significant voids may form in chalk within a time span of a few years (De Bruijn, 1983). However, this hazard appears to be negligible in comparison with the threat of collapse of either overburden soil or putty chalk, as the solutional effort is very unlikely to achieve its theoretical maximum.

Some areas of chalk contain numerous pipes, with loose fills of clay, sand and flint debris containing small voids. The pipes are conical or cylindrical, are normally a few metres across, and may reach depths of 30 m. They are essentially filled caves, and have formed where concentrated drainage has flowed into the chalk, normally near or just beneath outcrops of cover sediment and in areas of higher fracture density. Ground disturbance, by construction activity and drainage modification, caused a number of small collapses in a housing development on heavily piped chalk at Henley-on-Thames, and the housing units were mostly placed on raft foundations as a justifiable precaution (Edmonds, 1987). Most pipes have no surface expression, but their subsidence hazard is generally low, unless they are active; their main influence on civil engineering is often over quality control on cut-and-fill operations and the instability they create in cut slopes.

A high density of small conical pipes is comparable to pinnacled rockhead, and this is also common in the chalk beneath the feather-edge of overlying Tertiary sediments. As on limestone pinnacles, the uneven ground conditions make construction difficult, but the weaker chalk pinnacles are more easily removed. A pinnacled rockhead below sand, revealed along part of a motorway in Kent, was excavated until over 50% of the exposed material was firm chalk, and sites for bridge piles were probed 4 m below their bases to test for buried cavities (Higginbottom, 1966).

Sinkholes and depression may be common in the chalk surface, and Edmonds (1983) found densities in England ranging from less than $1/100 \text{ km}^2$ (mostly north of the Wash) to over $100/\text{km}^2$ in some small unit areas. They are very numerous in parts of Dorset, where they are mostly up to 50 m across and 10 m deep, with steep collapse profiles, all formed in the sands and clays which overlie the chalk (Sperling et al., 1977); in the main, these are typical subsidence sinkholes (see Chapter 3).

Isolated sinkhole collapses are reported from various parts of the chalk

Figure 4.1 A group of clay-filled pipes in the chalk, exposed in a quarry near Water End on the southern slope of the Chiltern Hills. Some almost cylindrical pipes are intersected obliquely by the quarry face, while the clay behind the person on the right fills a larger buried sinkhole. The top of the face is close to the original surface.

outcrop, including the well-known subsidence of the policeman's garden at Mickleham, south of London, in 1947 (Fagg, 1958). An unusual concentration of events was the 41 subsidences that occurred in the chalklands around Liège, Belgium, in the winter of 1966 (De Bruijn, 1983), all of which were due to loessic soils collapsing into chalk cavities at depths of 15–25 m. The Liège events may have been triggered by a small earthquake, or could have been due to a wet year and the temporarily high water table. As with limestone sinkholes, heavy rain is the commonest promoter of failures over chalk, though West and Dumbleton (1972) also note the effect of disturbance by construction traffic. A sinkhole into chalk undermined railway tracks behind a bridge abutment at Rainham, Kent (Toms, 1966); this occurred shortly after the bridge had been widened when site investigation boreholes had revealed no bad ground, though previously a cavity in the chalk had been found and filled. The placing of soakaway drains is also commonly responsible for subsidence activity, especially where the drains are in a thick cover of sand over the chalk.

Regardless of influence by construction traffic or by implanted drains, the greatest densities of chalk subsidences occur close to the boundary of overlying sands and clays, where a history of natural drainage input has created the most solutional cavities in the chalk (Figure 4.2). Most chalk sinkholes have formed beneath the feather-edge of, or very close to, outcrops of permeable cover deposits (Edmonds, 1983). They are virtually absent beneath impermeable clays, either Tertiary or glacial, or remote from impermeable boundaries, but are widespread beneath Tertiary, glacial or alluvial sands and

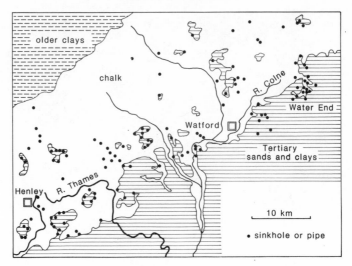

Figure 4.2 Distribution of known pipes and sinkholes in the chalk outcrop of part of the Chiltern Hills, northwest of London. Each dot represents either a single or a close group of sinkholes or clay-filled pipes, and there is a clear spatial relationship with the outcrop edge of the overlying Tertiary sands and clays (after Edmonds, 1983).

also along valley floors and in fold belts of fractured chalk (Edmonds *et al.*, 1987). A semi-quantitative hazard rating scheme has been devised by Edmonds *et al.* (1987) to facilitate land zoning and planning on chalk. Their main parameter is topography, whereby defined drainage channels and steep slopes create hazardous concentration of drainage compared with conditions on flat ground. Other parameters recognize increased sinkhole hazard due to adjacent cover rocks, a water table below rockhead, certain beds of weaker chalk, and past drainage routes.

Collapses in putty chalk

Chalk is especially susceptible to frost shattering, and heavily fractured rock may be described as rubble chalk, while the totally disintegrated, structureless material is known as putty chalk. Patches of putty chalk have been induced in test samples of highly porous Upper Chalk after fewer than ten cycles of freeze and thaw (Bell, 1977). Prolonged periglacial activity, notably in the extensive chalk outcrops just beyond the Devensian ice limits, has therefore produced putty chalk, typically to depths of 5 m and locally to depths of 30 m, in most natural chalk exposures, and also beneath thin drift deposits. The shattered chalk has then been prone to solifluction; chalk head is the widespread soliflucted putty chalk, and coombe rock is the same, with a variable degree of partial recementing by secondary calcite.

Putty chalk is structureless mélange, with irregular fragments in a remoulded matrix; Ward *et al.* (1968) outlined a chalk grading scheme,

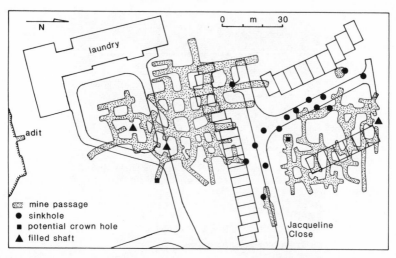

Figure 4.3 The collapse sinkholes and the underlying mines in the chalk at Jacqueline Close, Bury St Edmunds, Suffolk. The sinkholes are as recorded in 1977. Near the laundry building, the adit is a blocked mine entry in a small old quarry face. Many of the blind ends of the mine workings are blocked by fallen or stacked debris, so the mines are certainly more extensive than those that could be mapped (mine plans after Young, 1970).

comparable with the engineering grades of weathering, and their grade V material yields at bearing pressures below 200 kPa, exhibits significant creep, and has a standard penetration test N-value of less than 15. Putty chalk is usually sensitive, losing much of its strength when disturbed without drainage (Burland *et al.*, 1983). The liquid limit of the remoulded material is commonly close to the water content of many chalks. All putty chalk is mechanically disturbed, either by frost, solifluction or engineering works; the failure of excavated wet chalk, and its moderate thixotropy, are well known, as are cases of piles which can be driven almost endlessly unless given time to stabilize.

Sudden failures of putty chalk, creating significant sinkhole collapses, exhibit the properties of total liquefaction of the material. Small sinkholes are often recorded on construction sites, and many do not involve overburden soil failure. These can occur with no warning, but are related to disturbance by construction traffic, are often after periods of heavy rain, and depend on the presence of some buried cavity into which the liquefied debris can flow. Some of the natural chalk sinkholes, as described above, may involve putty chalk liquefaction in addition to surface soil failure.

A fine example of chalk liquefaction sinkholes is the suite that formed in a housing area at Bury St Edmunds in the late 1960s. Abandoned chalk mines underlie the site at depths of 10–12 m (Figure 4.3), and were not identified by an inadequate site investigation before the houses were built; stormwater was diverted into soakaway drains 6 m deep. Beneath only a thin soil, the chalk is weathered and frost-shattered; an adjacent hospital laundry was founded on

Figure 4.4 One of the collapses in Jacqueline Close, Bury St Edmunds, with the abandoned houses behind. The profile in the collapse reveals the tarmac on 30 cm of roadbase, with a broken drain on the right, on almost structureless putty chalk.

piles driven to 12 m, in some cases through small cavities. Starting before the last houses were completed, a series of collapses effectively destroyed the site. The failures were not true crown holes (see Chapter 5); the mines had been stable for over 50 years when only diffuse percolation drainage reached them, and the chalk appears to have failed rapidly over the whole depth (there are also separate incipient crown holes and infilled shafts which did not fail). Also, they were not normal subsidence sinkholes, as the failures involved clearly-defined cylindrical columns of chalk, and not just a soil cover. Concentrated drainage flow, between the soakaways and mines, had induced liquefaction of the already disintegrated putty chalk (Figure 4.5). It is possible that a pinnacled boundary may have existed between the putty chalk and the underlying chalk rock; pipes of putty chalk may then have localized zones prone to liquefaction at depth, but this control is speculative. Thirty years earlier some collapses over old mines in Norwich had in some cases involved clay-filled pipes which the mines had intersected (Edmonds, 1987); the pipe

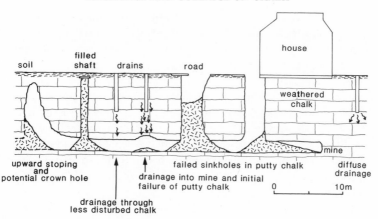

Figure 4.5 Cross-section to show the different features and stages of collapse at the chalk mines at Jacqueline Close, Bury St Edmunds.

sediments then failed, so this was a different mechanism from the failure of the weathered chalk at Bury St Edmunds. Some more recent collapses in Norwich include the one with the upended double-decker bus which achieved brief national fame on the television news programmes (Figure 1.1); again in this case there was an underlying mine, but there were no soakaway drains.

Circular sinkhole collapses have developed in cambered chalk slopes in Kent, and are probably due to liquefaction flow into underlying gulls (see Chapter 2). Near Troyes, in the chalklands of France, a sudden collapse breached a featureless cultivated hillside in the early 1970s. Over 16 m deep and 10 m across, its vertical walls of putty and rubble chalk are very similar to those in the Bury St Edmunds collapses. The role of drainage was demonstrated in Norfolk when a sinkhole destroyed a house as the water table was falling during a well pump test; the valley floor site had 10 m of sand overlying putty chalk, and failure appears to have originated in the latter. This collapse, the one at Troyes, and some other recent collapses near Norwich, have all occurred in areas with no known mines. It appears that these involve saturated putty chalk dropping into either solutional caves or networks of microcavities within the deeper chalk. As natural chalk cavities are likely to be small and with limited capability for underground sediment movement, the failures may have developed as periodic collapses with upward stoping, over longer periods than the apparently rapid failures at Bury St Edmunds.

Fortunately, putty chalk is largely a surface feature, and foundation piles for heavy structures will find stronger chalk, with N values greater than 40, normally at depths of 10–15 m. Furthermore, weathered chalk that is left undisturbed beneath a slab foundation may cause less settlement than indicated by SPT or laboratory tests (Burland *et al.*, 1983). Strengths of excavated chalks are greatly increased by admixed cement or bentonite (Lewis and Craney, 1966), but grouting is generally uneconomic for stabilization of in-situ putty chalk except where open cavities are identified and must be filled.

The major subsidence hazard is provided by the potential for liquefied putty chalk to flow into buried cavities, which may be old mines, solution caves or gulls. Chalk and flint mines are widespread and appear in a great variety of local styles (Edmonds *et al.*, 1987), but are always in hills above the water table. Solution cavities are more common under valley floors, and gulls occur in cambered hillsides, where clay underlying the chalk is exposed at the slope foot. Locating and sealing all cavities is commonly an unrealistic target, and any remaining hazard is always compounded by concentrated drainage from soakaways. A basic precaution is therefore to avoid the use of soakaway drains in chalk where site investigation reveals any possibility of old mines, sinkholes or cambering.

5 Mining subsidence

Total-extraction methods of mining remove all of the mineral present, thereby leaving an unsupported mine roof which is freely allowed to collapse. Consequently, ground subsidence is inevitable and extensive. The costs of surface subsidence damage, to both land and structures, are more than compensated by the increased mineral recovery when no support pillars are left in place, and also by the opportunity to mechanize longwall extraction faces in thin coal seams. The ground subsidence is predictable and rapid, in contrast to the random, lingering hazard of many old pillar-and-stall mines, and it is controllable to the extent that it can be locally reduced in some circumstances.

Underground mining of coal in Britain and much of Europe is dominated by mechanized longwall working, but this is not the only total-extraction technique. Much more of the coal mining in North America is achieved by pillar-and-stall working followed by pillar extraction on the retreat. And block caving of various metalliferous ore bodies can cause massive localized surface collapse. These latter methods are referred to again later in this chapter.

Longwall mining

A modern longwall coal face has a track-mounted coal cutter operating beneath the protection of skid-mounted hydraulic roof supports. The whole face advances as a slice of coal is sheared off on each pass of the cutter; the roof supports are then slid forward, and the roof is allowed to collapse into the area behind known as the goaf (Figure 5.1). Commonly the face is around 200 m across, and it advances to remove a panel of coal up to 2000 m long, over a period of about 2 years. Miners' access is through lengthening and supported (or retreating) gates down each side of the panel and connected to the main mine roadways which lie in stable pillars of untouched coal. Extracted seam thickness is usually 1–3 m, and this may lie at depths up to 600 m below ground.

Over the goaf, the roof breaks up and collapses almost immediately after the face supports are advanced. The roof rocks fracture, dilate and then close up as the movement zone migrates upward and the pressure of fallen overburden

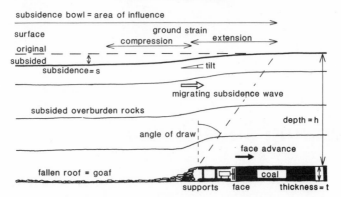

Figure 5.1 The main features of the subsidence over an advancing longwall face in a coal mine. For the purposes of the diagram, the coal seam has been drawn unrealistically thick for a modern mine.

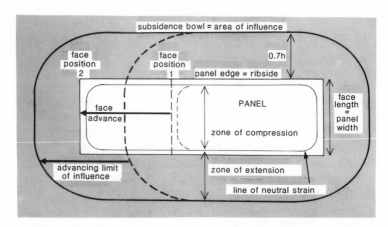

Figure 5.2 Plan of a coal mine panel with an advancing longwall face and an expanding area of influence. The position of the line of neutral strain, separating the zones of extension and compression on the surface, varies with respect to the panel ribside; this configuration applies where the ratio of panel width to seam depth (w/h) is between 0.6 and 1.4.

increases. Locally the roof rock movement may develop along joints or bedding planes, but the overall pattern is a gentle sag. The ground surface subsides into a bowl with an area much greater than that of the extracted panel. This subsidence bowl grows by migration of the subsidence wave which keeps pace over the advancing face; at the same time, static subsidence waves develop over the panel edges, or ribsides, behind the travelling wave (Figure 5.2).

The surface subsidence develops with various components of movement, each with its own influence on structures and engineering practice:

(i) Subsidence, or vertical displacement, is always less than the thickness of coal extracted, but it can accumulate in areas of multiple seam working.

Figure 5.3 Land flooding and a regraded road in a bowl of subsidence over an extracted longwall coal panel near Langley Mill, Nottinghamshire.

Parts of Hanley, in Staffordshire, subsided nearly 9 m in the period 1876–1966, but there is considerable local variation due to overlap of superimposed panels and conservation of shaft pillars (Hughes, 1981). Vertical subsidence may have little impact on building, but is often significant for river structures, where levée and weir heights and bridge clearances may be critical, and for lowland drainage and flood potential. Pennington Flash is a lake of 57 ha developed since 1900 in a subsidence bowl in Lancashire.

(ii) Tilt, or differential subsidence, is usually small over modern deep mines, as the length of the subsidence wave is comparable with the mining depth. In Hanley (Hughes, 1981) subsidence locally varied by up to 6 m over a distance of 500 m. Tilt is critical to tall structures, notably chimneys, and some industrial machinery, but its major effect is on canals, land drains, pipelines and sewers. Curvature, or differential tilt, affects long large buildings, and twisting movement on travelling subsidence waves can damage bridge structures.

(iii) Ground strain is developed as extension over the convex part of the subsidence wave, and compression over the concave section (Figure 5.1), with horizontal displacement reaching a maximum in the wave centre. Ultimate strain, the sum of extension and compression, is typically a few mm/m and causes the majority of building damage. Strains were highest over old shallow mines with steep subsidence waves, and long terraces of houses suffered worst, but generally less than 15% of buildings incur more than trivial damage over modern deep mines.

Figure 5.4 Destructive subsidence damage to a house on strip foundations at Hucknall in the Nottinghamshire coalfield. Ground movement was abnormally high because the house stood on fractured limestone broken by a fault zone to which the coal face had advanced and then stopped. The house was subsequently demolished and replaced by another with a raft foundation, though further mining beneath the site is unlikely.

The extent of mining subsidence and its subsequent damage is subject to three styles of influence. Mining factors are based on the pattern of working, and create largely predictable subsidence parameters. Site factors relate to local geology and any remains of old mines, and modify the subsidence in a much less predictable style. Structural factors then also dictate the scale of subsidence damage in terms of the movement tolerance of existing buildings and structures.

Subsidence prediction

Empirical formulae based on a mass of analysed data provide the most widespread means of subsidence prediction. Research and data collection developed first in Britain and Europe; the basic techniques were published in the National Coal Board's (now known as British Coal) Subsidence Engineers'

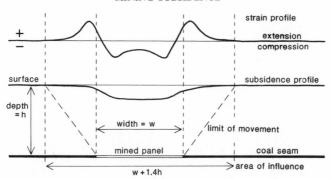

Figure 5.5 Typical profiles of subsidence and ground strain over a subcritical longwall panel in a coal mine.

Handbook (NCB, 1975), and were usefully reviewed by Shadbolt (1975, 1978). Parallel studies were also pursued in the Appalachian coalfields of the USA (Gray and Bruhn, 1984). The NCB predictions are generalized, and site factors, based on local data banks, need to be applied appropriately, especially in areas of strong overburden rocks (see below).

Extraction of a panel of coal creates a bowl of subsidence (or area of influence) whose size is determined by the angle of draw (Figure 5.1); this angle depends on overburden rock strength, and in Britain is taken as 35°, enlarging the area of influence by $0.7h$ (where h is mine depth) around the panel (Figure 5.2). Maximum subsidence is only achieved where $w > 1.4h$ (where w is panel width), and this is known as the critical width. An outer zone of extension surrounds an inner zone of compression, which is smaller than the panel area (Figure 5.2) except where $w/h < 0.5$. In supercritical panels ($w/h > 1.4$) an inner zone of nil residual strain is also formed.

A point over the panel is subjected first to extension, which then changes to compression shortly after the face passes beneath (Figure 5.2). The subsidence and strain profiles across the panel typically have the form shown in Figure 5.5. Variations do occur: over a wide panel with $w/h > 1.4$, the subsidence profile has a flat floor, and the strain decreases to zero across it; and over a narrow panel with $w/h < 0.5$, common in some deep modern mines, the central zone of reduced compression disappears and the transition from extension to compression moves outside the panel ribside (for further details see NCB, 1975).

Maximum values of subsidence, ground strain (both extension and compression) and tilt, over any worked panel, can be predicted because they can be related to w/h and t/h (where t is mined thickness) graphically as in Figure 5.6. Subsidence s is a predictable fraction of the thickness of coal extracted, and s/t varies with w/h (Figure 5.6); this fraction reaches a maximum only over a supercritical panel ($w/h > 1.4$), where the value of s/t is then known as the subsidence factor for that particular coalfield. Maximum horizontal displacement approximates to $s/6$. Maximum extension and compression, and their

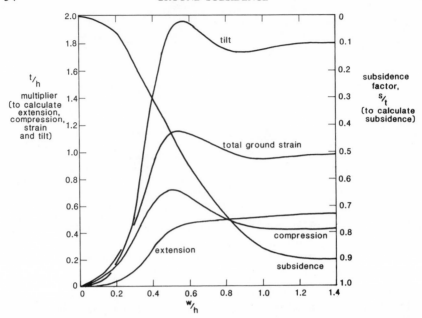

Figure 5.6 Graphical determination of subsidence parameters over extracted longwall panels. For a known value of w/h, unitless numbers can be read from the graphs. Subsidence is the value on the right scale, multiplied by t; and the other parameters are the values on the left scale multiplied by t/h; where h is seam depth, t is mined thickness, and w is panel width (after NCB, 1975).

total which is ultimate strain, relate to both w/h and t/h. Figure 5.6 is used to derive unitless values for a given w/h, which are then multiplied by t/h, to give the strains (so that, for example, over a panel where $w/h = 0.4$, extension $=$ $0.3\,t/h$, and if $t/h = 0.005$, extension $= 0.0015$ or 1.5 mm/m). Maximum tilt is similarly calculated off Figure 5.6. For supercritical panels (where $w/h > 1.4$) there is no further change with respect to w/h in any of these values.

The travelling subsidence wave, over the advancing face, is roughly similar to the lateral transverse waves (Wardell, 1957). Its development of subsidence and strains over the coal face is shown in Figure 5.7, and its migration rate is the same as the face advance. Practically all the surface movement occurs within the wavelength of about $1.4\,h$, and this means that a single point over a modern deep mine panel is usually affected for about nine months. Residual subsidence after that time normally occurs for perhaps another year, but is negligibly small.

The data and graphs within these pages only allow prediction of maximum subsidence values, appropriate for outline or rather conservative engineering applications. Prediction of more detailed subsidence profiles, and also consideration of points on the margins of the subsidence bowl, not subject to maximum movements, are possible with the Subsidence Engineers' Handbook (NCB, 1975).

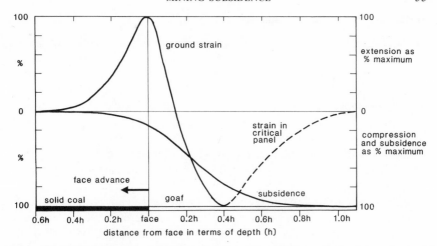

Figure 5.7 Development of the subsidence wave and strain profile at a single point as a longwall face passes beneath. Note that the tail of the strain profile is modified to leave residual compression if the panel is subcritical (after NCB, 1975).

Subsidence predictions based on these empirical methods provide values which are generally reliable to $\pm\,10\%$ (except in some cases where abnormal variations are imposed by site factors; these include the presence of competent rocks, faults, seam dip, surface slope and multiseam working, as described below). In practical terms this is normally adequate, but predictions are rarely reliable more than about three years ahead (Buist and Jones, 1978) due to unavoidable changes in panel layout that can only be made as mine development reveals geological detail.

The classic case of mining subsidence prediction concerns Duisburg harbour, Germany's main inland port on the Rhine. Channel improvements had caused water levels to fall 2.4 m in 100 years, and the 45 km of wharfage at Duisburg required comparable lowering. Between 1956 and 1968 this was achieved by carefully planned mining subsidence (Legget, 1972). Twelve million tons of coal were mined from three seams, totalling 3.8 m thick at depths of 80–500 m, creating a subsidence bowl 3 km across in which most of the harbour was lowered by 1–2 m. Panels were designed to achieve the required subsidence, and movement had to stay close to predicted values; a major lock temporarily had 360 mm of differential subsidence, but finished 480 mm lower than it started—against a prediction of 500 mm. There were no major sandstones to promote erratic subsidence, but seam faulting curtailed one small zone of mining, leaving above it less subsidence than had been designed. A minor disaster occurred when a string of railway wagons, left unbraked on level track, rolled away when tilted by the migrating subsidence wave. Otherwise, harbour operations continued uninterrupted.

C

Site factors and abnormal subsidence

The National Coal Board prediction system is used in coalfields worldwide, but it does frequently require adjustment, especially where the overburden rocks are significantly different from the British national average. Even within Britain, some coalfields have their own regional data graphs to improve upon the national averages and allow more accurate predictions. Subsidence damage over American coal mines is often severe, and Gray and Bruhn (1984) emphasized the danger of using NCB data in the USA as conservative predictions were then achieved. This is due to different effects that the pillar-and-stall retreat mining may have from the longwall method, the higher sandstone–shale ratio which increases surface strain, the higher strains created in undisturbed ground in single seam coalfields, and the dominance of easily damaged basements in American houses. Data have now accumulated to the extent that many of these variations can be assessed and, with adequate knowledge of local site conditions, prediction reliability can often be maintained.

Multiple seam mining merely accumulates subsidence effects, but subsidence over first seam working in virgin ground does show some contrast with that from subsequent mining through disturbed ground, which is the normal situation to which NCB predictions can be applied. Mining through undisturbed ground promotes a reduced subsidence factor (Figure 5.8) of 0.8 instead of 0.9 (Shadbolt, 1978), though generally with higher amounts of residual subsidence, and induces much higher maximum ground strains in the Illinois coalfields (Hood *et al.*, 1983). Mining beneath old partial-extraction pillar-and-stall workings may cause very erratic subsidence due to induced pillar failures (see Chapter 6), has generated minor earthquakes in the Stoke-

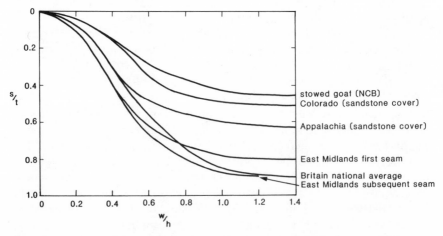

Figure 5.8 Variation of the subsidence factor, for given values of w/h, due to multiple seam working, sandstone overburden and stowing of waste into the goaf (after NCB, 1975; Shadbolt, 1978; Dunrud, 1984; Gray and Bruhn 1984).

on-Trent coalfield (Westbrook *et al.*, 1980), and has been found to have a higher angle of draw at some sites (Priest and Orchard, 1958).

The dip of a coal seam and its worked panel has a significant and well-documented effect on the ground subsidence. The subsidence is developed not vertically above the seam, but for dips up to about 30°, is developed normal to the bedding, thereby displacing the area of influence towards the downdip side and also distorting the profile. A dip of 10° increases the angle of draw (measured from vertical) by 12° and increases tensile strain by 30% on the downdip side, with comparable reductions updip, though neither relationship is linear (NCB, 1975).

Steep ground slopes distort subsidence profiles, especially where lateral displacement is enhanced where draw into a bowl of subsidence coincides with the downslope, and mining subsidence can even promote landsliding. On valley sides in the South Wales coalfield, strain profiles are displaced downslope and also vary with the direction of mining relative to the slope (Franks, 1985). Where mining advances in the downslope direction, extension is reduced and compression is unchanged, while mining in an upslope direction decreases compression and greatly increases extension.

Faults commonly disturb the subsidence profile, causing local concentrations of strain and even ground stepping (Hellewell, 1988). Their behaviour is not accurately predictable, and faults in sandstone outcrops in England react abnormally in 40% of cases (Shadbolt, 1978). A survey in the Yorkshire coalfield (Lee, 1966) found all major and many minor faults causing some ground stepping, which was most frequent where mining was beneath the hade of the fault. Most fault steps developed over the panel and 0.2 h in from the ribside or face, especially where the fault was parallel to a long advancing face,

Figure 5.9 A road in the Yorkshire coalfield is deformed across a step nearly half a metre high, where a small fault has localized ground movement on the edge of a subsidence bowl over an extracted longwall panel.

and the ground step was commonly around 30% of the maximum subsidence induced on the panel.

The lithology of the overburden rocks has a significant influence on the subsidence parameters. The NCB predictions are based on average British site conditions, which are dominated by weak shale and mudstone, and the subsidence profile is distorted by any increase in the proportion of competent sandstone or limestone between the mine and the ground surface. In Britain, the subsidence factor (s/t) is usually taken as 0.9 (NCB, 1975), but American coalfields with their more competent cover rocks have typical values of 0.7 or less (Figure 5.8). The Salina coalfield in Utah has subsidence factors of 0.9 over shale, but 0.45 over strong sandstone (Dunrud, 1984), and the first faces in England's new Selby coalfield have produced less than predicted subsidence due to the Magnesian Limestone in the cover (Pyne and Randon, 1986). Data gathered in the Appalachian coalfield (Tandanand and Powell, 1984) give the rough guide that subsidence decreases from the NCB prediction by a percentage equal to that of sandstone in the cover rock sequence, but the Australian Kemira coalfield has subsidence factors of 0.6–0.8 with 70% sandstone in the cover (Wood and Renfrey, 1975).

Angle of draw also changes with overburden lithology. In Britain it ranges 25–35° (with NCB predictions using the conservative 35°), but loose cover sands in Holland give values up to 45°. Coalfields with more sandstone give angles of 10–28° in the USA (Gray and Bruhn, 1984) and 13–28° in India (Kumar et al., 1973), and some Illinois coalfields have recorded transverse angles of draw of 43° along with longitudinal angles of 17°, probably due to anisotropy in the overburden (Hood et al., 1983).

Maximum strain values also increase on the steeper subsidence waves formed in more competent rocks. Strains over a New Mexico coal mine were 50–100% higher than NCB predictions, due not only to the strong overlying sandstone but also to the first seam working and steep surface slopes (Gentry and Able, 1978).

There is no doubt that overburden lithology does exercise a major control on mining subsidence profiles. As the complexity of lithological variation makes generalized correlation difficult, local data banks to substitute in and improve upon the empirical NCB prediction methods must further gain in importance.

Competent rocks at outcrop cause even more erratic mining subsidence, as movement is concentrated in fissures, and the intervening blocks act as monoliths. The Sherwood Sandstone and Magnesian Limestone, overlying much of England's concealed coalfields, are prone to this behaviour, and account for subsidence being abnormal, with respect to NCB predictions, in around 25% of cases on their outcrops, as opposed to less than 10% on shale outcrops within the Coal Measures (Shadbolt, 1978). From Colorado, Dunrud (1976) cites dramatic examples both of tension fractures opening by 300 mm and of rock slabs uplifted in compression bulges in a sandstone outcrop

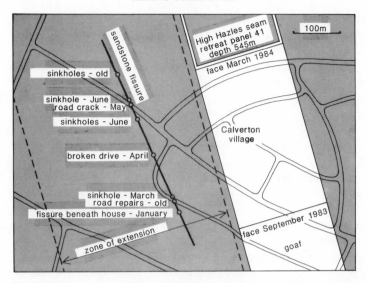

Figure 5.10 Subsidence features which developed in 1984 along a fissure in soil-covered sandstone within the zone of extension obliquely above an advancing longwall panel at the Calverton colliery in Nottinghamshire. The subsidence sinkholes formed when the soil was washed down into the opening fissure. Old features along the same fissure date from earlier movements when a parallel panel was extracted from a higher seam.

270 m above a mine. Where a fissure is masked by soil, its opening may express itself just by subsidence sinkholes within the soil (Dearman, *et al.*, 1982) and by localized structural damage. Figure 5.10 shows subsidence features over a Nottinghamshire coal mine where a fissure in sandstone was in the most sensitive position, nearly parallel to the panel ribside and in the heart of the extension zone; tensile strain across the fissure reached 0.0004 (a low value due to the great depth of the mine) and the width of opening suggested it was taking strain from a zone of ground about 50 m wide.

Notoriously severe subsidence is experienced over gulls close to scarp edges in the Magnesian Limestone of Nottinghamshire. These fissures, originally opened by cambering displacement above underlying plastic clays, may be over a metre wide and open or rubble filled to depths of 20 m, and yet lie hidden beneath the soil blanket (Buist and Jones, 1978). Differential movement across them, and settlement of their fill during subsidence, account for frequent cases of abnormal and largely unpredictable structural damage on the limestone outcrop.

Towards improved subsidence prediction

The widely used NCB methods of predicting coal mining subsidence have proved immensely valuable in engineering design and planning, but they are only generalized and simplified approximations. One limitation stems from the two-dimensional approach, which does not easily allow for twist and shear, and there is a need for more three-dimensional monitoring of subsidence and strain data (Chen and Peng, 1986).

Of the mathematical approaches to subsidence, a profile function has been found to provide better predictions of curvature and strain than either the NCB data or more problematic, though more flexible, influence functions (Hood et al., 1983). However, this technique still depends on constants derived from earlier locally recorded data, and it has been found less successful where more than a few seams are worked or mining depths exceed 400 m (Brauner, 1973).

No mathematical prediction can be better than the assumptions made and the parameters introduced within it, and this commonly brings its true value back to the generalizations of the empirical methods. Mathematical description of complex variations within a rock sequence is not easily achieved, and most mathematical approaches ignore the effects of cover rock lithology (Chen and Peng, 1986), whose influence on subsidence is clearly proven.

Computer models can employ influence functions which consider a mass of data and predict subsidence by summing the effects from all mined and unmined sub-areas. They benefit from their scope to adjust to huge banks of recorded data, and can better appraise the curved three-dimensional pattern of the subsidence. To predict subsidence where continuously-welded rail track has to be maintained for high-speed trains, refined progams have been developed for British Rail (Burton, 1978, 1985) but these exhibit a complexity which may render them unwarranted at less critical sites.

Pillar-and-stall mining with total extraction

As an alternative to the longwall layout, mining can cut conventional stall headings to take 50% of the coal from a panel in the first instance, and then remove the pillars which constitute the other 50% on the retreat, allowing the roof to collapse beyond. This method is used in about half of North American coal mines, usually as a room-and-pillar working with square pillars (Gray and Bruhn, 1984), though bord-and-pillar working with elongate pillars (Sladen et al., 1984) is a common variant. The end result is similar to longwall mining, in that total extraction is achieved from the panel, while coal is left in place to form roadway pillars, and the systematic removal of an area of pillars creates a migrating subsidence wave.

In broad terms the ground subsidence produced by the pillar retreat is comparable with that over a longwall panel of equal dimensions (Dunrud, 1984); consequently the empirical methods of qualified subsidence prediction, outlined above, remain appropriate. However, the subsidence ratio may be reduced by unextracted pillar remnants or stumps, and experience in the eastern USA suggests that the angle of draw is often a few degrees greater than over a longwall panel (Gray and Bruhn, 1984). Additionally, the patterns of roof stressing and subsequent breakdown may exhibit contrasts between the two mining methods, but these would appear to be less than the variations due to lithology and other site factors. Ground movements also tend to exhibit more frequent local irregularity than over longwall mines; where difficult mining conditions cause a pillar to be abandoned in place, full roof collapse is prevented and a hump is created in the subsidence profile (Speck and Bruhn, 1983).

Subsidence over deep caved mines

Metalliferous ore bodies occur in a considerable variety of shapes, and numerous mining techniques have evolved to ensure their efficient abstraction. Especially in steep or vertical ore bodies, the economic benefits of total-extraction subsidence techniques are significant; ground subsidence is then inevitable and may involve major vertical movements, though its lateral extent is only comparable to the usually quite limited dimensions of the ore body, and bears no similarity to the vast subsidence areas over coal mines.

A buried ore body may be worked by block caving, or sublevel caving, where the ore is broken in situ and then tapped off from below, with the collapsed roof descending on top of the reducing pile of broken ore. Alternatively it may be mined from the top, when a slice is taken, in a sequence comparable to bord-and-pillar working followed by pillar removal, with the process then repeated consecutively on the next slices below, working beneath the timbered fallen roof. In either case, roof failure develops through progressive caving, or upward stoping, unless the ore body dimensions allow formation of stable arches.

Where overburden is relatively thin, the roof caving may work directly to the surface, creating, in effect, a very large crown hole. Ultimately this develops into a steep-sided surface pit, floored by broken overburden and resembling a quarry, except that it has been excavated from below. Subsidence of the conglomerate caprock over an Arizona copper mine has created a pit 900 m across and 150 m deep (Hatheway, 1968); the pit is ringed by vertical tension scarps and is formed by progressive collapse into a central vertical pipe where the ore was totally removed at depth.

Over deeper caved mines the patterns of overburden failure are less predictable. Caving of the inclined copper ore body at Mufulira, in Zambia, caused overburden breakdown from a depth of 500 m, which ultimately caused localized surface fissuring beneath a mine tailings dump (Sandy et al., 1976). In the tailings a sinkhole 280 m across and 15 m deep developed within 15 minutes, though in this case the disastrous inrush to the mine below was more serious than the ground subsidence damage.

Unlike many coal mines, metalliferous mines are rarely in areas of high population density, and measures to avert subsidence are therefore less applicable. Expensive hydraulic sand backfilling, or partial-extraction support methods of mining, are not justified merely to preserve undeveloped land. The

Figure 5.11 Subsidence of an old sea wall above the Hodbarrow Mine in Cumbria. A large ore body of hematite iron ore was totally extracted from beneath 15 m of limestone and 60 m of drift. Surface subsidence was inevitable, and the sea was kept out by another outer barrier. Since the mine closed, the sea has reinvaded the subsided area to create a lagoon which is now a nature reserve (Photo: British Geological Survey, © NERC).

scale of subsidence may sterilize the land, but the areas involved are small and are often suitable for mine-waste dumping. Some major subsidences were created over the iron ore mines, now closed, in Cumbria (Holland, 1962). The largest subsidence bowl, spanning nearly 20 ha and more than 20 m deep, has since been invaded by the sea. Others in low ground remain with lakes in steep-sided depressions, while subsidence under hillsides created sharp-edged troughs of chaotically broken, collapsed ground.

Mining precautions to reduce subsidence

High levels of confidence are achieved by modern subsidence predictions over coal mines which extract only laterally extensive, thin seams. Consequently mine planning is refined to the stage where it can not only maximize coal extraction but can also minimize environmental disturbance. Britain's new Selby coalfield underlies a flood-prone lowland, and it is completely practicable to impose strict limits on the allowable ground subsidence (Pyne and Randon, 1986). Mine design can also reduce subsidence damage by ensuring that sensitive structures lie over the centres of wide panels, where residual strain is minimal, or parallel to the advancing face, to avoid twisting movement (Wardell, 1957), though the ideal scheme is often restrained by the available choice of practical panel layouts.

The empirical predictions of the NCB (1975) even extend to assessment of structural damage due to the subsidence. In British coalfields, individual panels are costed for the overall subsidence damage that they are likely to promote. Some urban areas may have subsidence costs rising to £20 per ton of coal, or 30% of production costs, compared to the norm of less than £5 per ton. Potential high cost areas are designated Subsidence Sensitive Zones, and are not mined. Effectively this means that urban areas are rarely undermined in Britain, though this was not the case in the past and may not be in the future if energy costs rise.

Aside from this broad approach, individual structures may warrant special consideration. The options then are to move the mining or move the structure. The frame-construction houses prevalent in America can be removed before mining and replaced afterwards (Chen and Peng, 1986), while, in England, 14 km of new railway now takes high-speed trains around the Selby coalfield (Pyne and Randon, 1986), and, in 1975, a complete church was moved 700 m to avoid an impending subsidence bowl at Most in Czechoslovakia.

A mine can be designed to leave in place pillars of coal to support sensitive structures such as churches, dams or power stations. However, these pillars have to be wide enough to allow for the angle of draw, so sterilizing large tonnages of coal reserves in deep mines, and they also complicate the layout of longwall panels around them. It is much easier and cheaper to leave partial support pillars in American pillar-and-stall mines, where zones of pillars are simply left standing on the retreat. Single houses over shallow mines in the Pittsburgh coalfield are effectively stablilized over pillars designed with a 5 m margin in plan, plus a 15° angle of draw, and with 50% extraction within the pillar (Gray and Meyers, 1970). Stowing of waste rock into the goaf, before the roof collapses into it, can reduce subsidence by up to half (Figure 5.8), but this is a difficult and costly procedure of limited application in mechanized mines.

Subsidence is reduced over narrow panels, and where $w/h = 0.25$ is little more than 10% of its maximum (Figure 5.8). Longwall panels were often supercritical in old shallow mines, so creating maximum subsidence and damage, and shortwall working on faces only 50 m long was sometimes introduced. This is no longer necessary where the greater depths in modern mines favourably reduce the w/h ratio even in optimum longwall panels. Reductions in ground strain can be achieved by a form of partial extraction with narrow panels between wide pillars. A seam 2.5 m thick was mined beneath an English city, and 50% extraction with panels and pillars each only $0.25 h$ in width ensured only very slight ground movement (Shadbolt, 1978).

The pattern of narrow panels and equal pillars has the benefit of overlapping strain profiles, where tensile strain from one panel partly cancels out compressive strain from its neighbour. This principle is continued into harmonious mining where panels in separate seams are located so that their strain profiles are self-eliminating. Though practised in some European coalfields, the necessary mine layout is not always feasible, and difficulties also occur where delays on one face disturb the timing of balanced ground movements. A version of harmonious working is the stepped face layout, where a wide panel is advanced by two half-length faces, one lagged behind the other by a distance ideally around $0.4\,h$, to reduce strain in the zone of overlapping influence. A Nottingham church was successfully undermined with negligible damage using this technique in 1956 (Priest and Orchard, 1958). A panel 550 m wide was worked in two faces stepped 80 m apart at a depth of 195 m, and 0.8 m of coal was taken, causing subsidence of 0.64 m. But the total ground strain was only 0.002, which the church fabric could withstand, whereas an intolerable strain of 0.005 would have been imposed by a single face.

Structural precautions in subsidence areas

Construction planning rarely has the opportunity to avoid the severe subsidence areas, even where predictions are available (Bell, 1987b). However, undermined outcrops of competent rock commonly have movement concentrated along fissure lines; a single house within a terrace may suffer repeated damage as coal faces work beneath it, while its neighbours remain unscathed. Ideally, fissures and faults should be avoided by new structures, but their presence is effectively masked by a soil cover. Some fissures may be identified beneath a soil by tonal contrasts on air photographs, but the success rate is low, and the best data may come from the damage records of earlier buildings on a site being redeveloped. Large fissures or gulls only revealed during excavation are best filled with sub-base or lean concrete, or provided with a slab cap (Buist and Jones, 1978).

As subsidence sites cannot be avoided, structures have to be designed to live with the ground movements, and must either have complete flexibility or lie as small rigid units. Raft foundations are normal for buildings, and smooth-based reinforced concrete slabs should not exceed 20 m in length, unless they are considerably thickened as appropriate for high-rise structures (NCB, 1975). The slab is best placed on a loose granular base 150 mm thick, preferably with an interlayer of polythene, to let the ground strain independently beneath it, but loose sand cannot be used on a fissured bedrock. Substructures, either basements or piles, should be avoided as they are too sensitive to ground movement, and potentially severe compression may be alleviated by an encircling trench filled with polystyrene blocks.

Buildings too large to occupy single rigid slabs require design for flexibility. For example, hinged frames are preferable while brick arches should be avoided; numerous features of appropriate structural design are reviewed by Geddes (1970) and Singh (1979). The classic example of design specifically for subsidence-prone areas is the CLASP structure, whose concept originated with schools being built on the English coalfields (Lacey and Swain, 1957; Bell, S.E., 1978). A floor of unit slabs, no larger than $30\,m^2$, hinged by the rebars, supports a pin-jointed steel frame, braced by internal springs, and fitted with

Figure 5.12 Concrete raft supporting a house at Calverton in the Nottinghamshire coalfield. The trench has exposed a fissure in the sandstone bedrock, beneath the soil and weathered zone. This fissure is the one shown in Figure 5.10, which opened in tension due to mining subsidence; though the fissure caused surface damage elsewhere, the rafted house was unharmed.

flexible panels, so that a strain of 0.003 can be tolerated even by extensive building complexes.

New roads through subsidence areas should have a flexible macadam surface on a thick granular sub-base (Buist and Jones, 1978). Bridges require special attention, and subsidence precautions often add approximately 20% to their cost (Davies and Smith, 1978). The major threat to bridges comes from compression between approach embankments, and movement joints, preferably rubber or bitumen filled, within the structure should generally permit a total strain of 0.005 unless reliable local prediction dictates otherwise. Articulated structures, with hinged piers, roller bearing span support and perhaps three-point slab bearings and inbuilt jacking points, can accommodate even severe movements. Alternatively, flexible units were found better than articulation for a major motorway interchange in an area of future mining near Glasgow (Murray, 1984). Bridge foundations should have

Figure 5.13 Jacking points built into the brick clad concrete structure of a railway bridge to allow trackbed raising and relevelling in the event of anticipated subsidence.

spread footings, with a layer of sand, or alternatively low-viscosity bitumen, to absorb ground strain below the concrete pad.

The slope of a subsidence profile can severely affect water flow, and sewers may warrant construction with steeper gradients to allow for imposed tilts which can easily reach 1 in 100. Pipelines require flexible joints to cope with varied ground movements, and appropriate specifications are defined by the NCB (1975). There is also a case for more care over land drainage design in low-lying areas, such as over the new Selby and Vale of Belvoir coalfields, where even a shallow subsidence bowl can have a serious impact.

Preventive works on existing structures

Where existing structures are to be undermined, subsidence damage can sometimes be reduced by appropriate preventive measures. Foundations without a raft can be protected by an encircling trench taken to below foundation level and filled with compressible material such as polystyrene. Costs are around £15 per metre run of a trench 1.2 m deep, and this should halve the effective compression on the buildings (Shadbolt, 1975). In the Appalachian coalfields, most damage is just to basements, with their brittle concrete block walls beneath a more flexible timber-frame house (Bruhn *et al.*, 1983); deep trenches are the only effective strain protection, but are only partly successful.

Larger buildings suffer less damage when cut into smaller units. This is a simple preventive measure where corridor units linking a building complex

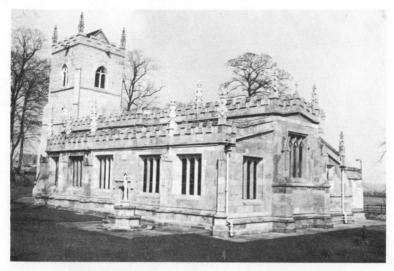

Figure 5.14 St Wilfrids church at Hickleton, in the Yorkshire coalfield, restored and underpinned to prevent any damage during future planned mining subsidence.

can have expanding joints inserted. Long terraces of brick houses are prone to severe damage from a migrating subsidence wave; cutting them into units by removing about every fifth house may seem drastic but can be worthwhile. When a Derbyshire village was undermined with a 1.5 m seam extraction, occupier resistance precluded any preventive measures, and most of the 100 houses were then so severely damaged that they were subsequently demolished (Smedley, 1977).

The brittle and irreplaceable fabric of England's churches is rarely undermined. But Hickleton church, in the Yorkshire coalfield, was recently underpinned on a grand scale as it had suffered subsidence damage in the past and could be undermined in the future (Dadson, 1984). The church stands on heavily fissured Magnesian Limestone close to the escarpment edge, and earlier stonework cracking showed where it straddled a significant fissure (which later proved to be a small fault). The underpinning treated the church as a bridge slab weighing 3600 t. Mass concrete beams under each wall bear onto a rectangular ferroconcrete frame entirely supported on jackable columns founded on three pad footings which impose a maximum load of 600 kPa on unweathered bedrock (Figure 5.15). Costs of the operation were £800 000, justified by a combination of the coal reserves made available and the social value of the ancient church.

Preventive or remedial works on bridges can prove costly, as experience along the M1 motorway across the Yorks–Derbys–Notts coalfield has shown. The motorway was built around 1964, when coal was in decline before rising energy costs increased the extent of mining, and the bridges were generally designed to tolerate ground strain of 0.003. Recent mining subsidence has

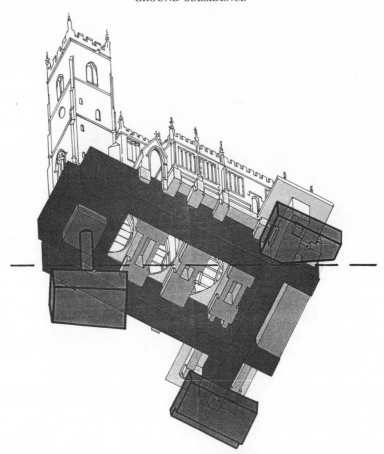

Figure 5.15 Worm's eye view of the massive concrete underbeams which now entirely support the church at Hickleton in the Yorkshire coalfield. The beams bear solely onto three pad footings via jacking points which can be adjusted to accommodate any future mining subsidence. (Drawing by kind courtesy of Hill Rowe Partnership Chartered Architects and Ove Arup and Partners Consulting Engineers).

surpassed this strain, and many bridges have required emergency works. Temporary Bailey or Hamilton bridges have been built on the carriageways to completely span some underpass bridges and remove from them the loading by heavy lorries while they have been overstressed by subsidence movement (Parkinson, 1981). Many overbridges have had temporary supports while a subsidence wave passes. One at Annesley was monitored when 2 m of coal was taken from a panel 220 m wide directly under the bridge at a depth of 475 m. Beneath each carriageway span the ground first extended but then showed residual compression (Figure 5.17), and total strain was close to the predicted 0.0045. Finally, 200 mm was cut from the eastern span and abutment, and the central pier was allowed to rotate. Further north on the motorway, an

Figure 5.16 Temporary structures for an overbridge occupying the fast lanes of the M1 motorway, where it passes over a working mine panel in the Nottinghamshire coalfield. Part of the load of the bridge was taken on sets of jacks mounted on the temporary frames; this provided precautionary support and also permitted monitoring of any twisting strain on the bridge. The horizontal strain record of the same bridge is shown in Figure 5.17.

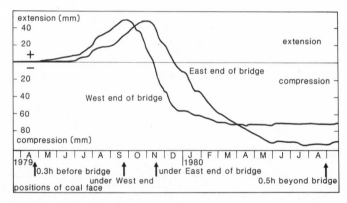

Figure 5.17 Observed ground strains at the two ends of an undermined bridge over the M1 motorway at Annesley in Nottinghamshire. The positions of the coal face which advanced beneath the bridge, are shown with respect to the depth h of 475 m. The strains were measured each week between studs placed in the bridge on each side of the expansion joints (strain data after Nottinghamshire County Council).

overbridge near Wakefield had its suspended span removed during subsidence, and replaced only when movement ceased and a total of 150 mm had been cut from the spans and abutments to accommodate residual compression (Jones and Spencer, 1978): total cost of these works, including a temporary Bailey bridge, came to £94 000 in 1978 (£195 000 at 1987 values).

Buried pipelines commonly demand preventive works where affected by subsidence, and are safest when exposed, fitted with flexible joints and supported on adjustable chocks during the movement. Sewer lines may then need rebuilding on permanent supports across a subsidence bowl (Smedley, 1977), but pressure lines can be reburied (Priest and Orchard, 1958). Land drains may require pump installations where gradient is lost, and stream courses can be lined with plastic sheeting where dewatering is threatened by ground extension (Chen and Peng, 1986). Low-gradient trunk streams and rivers may also require additional levée protection; some of the levées along the River Trent, and now also the Yorkshire Ouse, have been raised to accommodate mining subsidence.

Subsidence compensations and legal constraints

In Britain the coal industry is nationalized, and all the coal, except that in a few small areas such as the Forest of Dean, is owned in the ground by British Coal (formerly the National Coal Board). They enjoy an automatic right to mine the coal, but are similarly bound to pay compensation for surface damage. Subsidence costs currently average around £1–5 per ton of coal mined, and this is included in any mine budget. British Coal are also obliged to reveal all requested information on mining past, present and future (within the reasonable targets of advance planning); mine plans are open to inspection, all subsidence prediction data are available, and consultation between engineers and British Coal's subsidence specialists is beneficial to all parties. The legal obligations of British Coal are defined in a 1957 Act of Parliament, with amendments in a 1975 Act, and further details are enshrined in the current Code of Practice: the legal situation is usefully summarized by Reeves (1984).

Precautionary works on new structures, and appropriate designs of the same, can be recommended by British Coal, but prohibition of unsuitable structures in mining areas can only be enforced by the planning authority, who must in turn consult British Coal. Failure on the part of a developer to act on the British Coal advice may restrict the extent of subsequently claimable subsidence damage compensation. However, the costs of precautionary measures are normally carried by the developer; British Coal have no legal liability to pay them, even when they have stipulated the measures, and regard them as being natural costs associated with development of land which happens to overlie extractable coal. This issue is disputed by some local authorities, and the Subsidence Compensation Review Committee (1984) has recommended that costs should be shared between the developer and the mining authority. Even though not yet obligated, British Coal does agree, by negotiation, to bear some of the costs of some modifications to foundations and also of some works with wider implications such as raising the height of river levées.

Under the 1957 Act, British Coal has the power to carry out preventive works on existing structures due to be undermined, and, where the work is not done by their own contractors, they do bear the costs. The same Act then states that when the subsidence movement does eventually occur, British Coal must repair all damaged structures, or pay for the work to be done, or make a depreciation payment if the repair cost would be greater than the value lost. Full compensation is paid, except when structural precautions were advised but not carried out, when compensation is limited to the lesser cost of the remedial works which could have been necessary if the precautions had been made; also, repairs are only required to return the structure to its original use, and not to identically restore. The other scope for variation is where the subsidence can be shown to be, partly or wholly, due to causes other than mining. As an example, a Nottingham terrace house suffered severe damage but was then found to have been built over the edge of a forgotten backfilled quarry; after negotiation by its subsidence engineers, British Coal paid an agreed fraction of the compensation claim on the basis that the mining subsidence wave might have disturbed the old quarry fill.

Compensation and legal obligations concerning mining subsidence are totally different in America where mines of coal and other minerals are privately owned; the same applies in most other countries, and also to the few mines of other than coal in Britain. The mineral rights originally purchased for a given block of land include the right to subside the ground (Gray and Meyers, 1970), and should therefore incorporate agreement on compensation for any damage incurred. This situation reduces the need for the legislation on precautionary works established for British coal mining. However, the mine operator in the USA is still constrained, by laws enacted in 1977, to maintain subsidence control, and is normally prohibited from undermining perennial streams, reservoirs with capacities greater than $24\,600\,\text{m}^3$, certain public facilities such as schools and most urban areas. He must also provide subsidence predictions and a mitigation plan, and must remedy or mitigate subsidence damage to structures and to renewable resources, including aquifers (which may have been dewatered), forest and farmland (Chen and Peng, 1986). Some American mine operators offer home owners the option to purchase 50% of their underlying coal to form a partially mined pillar, in exchange for a guarantee of stability and repair of any damage (Gray and Meyers, 1970). This is rarely economic for a single house, except over a very shallow mine demanding only a small pillar, but may be practicable for small communities. Without this sort of protection, compensation agreements form the only feasible approach to the inevitable subsidence damage that stems from modern total-extraction mining.

6 Failure of old mineral workings

Underground mining, both today and in the past, is a major concern in many parts of the world, where its scale and extent can surprise anyone not directly involved in the industry. The removal of mineral ores automatically creates underground voids, which are normally neither feasible nor economic to stow with waste, and which therefore become problems to surface engineers when subsidence threatens.

Styles of mining vary enormously. They include total-extraction methods, where subsidence is inevitable, but which leads to a new stability; these are dominated by modern coal mining, as explained in Chapter 5. The alternative method of partial extraction leaves some mineral behind to act as roof support; mine economics then demand maximum extraction and minimal support, which, especially in past eras of short-sighted planning, created ground of marginal stability with a long-term threat of subsequent failure.

A large proportion of mineral ore bodies are planar sheets—either sedimentary bedded units, or fault-guided vein infills—and their mining style then relates largely to their dip. Vertical or steep veins, especially in strong country rock, are worked to leave deep open voids—or stopes—with minimal support of stable walls. Normally a few metres wide, stopes may be hundreds of metres long and deep. Lower angles of dip require progressively more support for the hanging wall or roof, usually in the form of remnant pillars of mineral or perhaps timbers, whose eventual failure will delay ground subsidence for maybe hundreds of years.

Pillar-and-stall mining is the basic method of partial extraction of level or low-dip ore bodies, notably coal seams. The geometry of the mines varies considerably along with the terminology. Bord-and-pillar (as still used in many American coal mines) and post-and-stall both leave elongate pillars, while stoop-and-room (similar to pillar-and-stall in many modern stone mines) and room-and-pillar both leave square pillars; Figures 6.1 and 6.2 show typical patterns and further examples are given by Piggott and Eynon (1978) and Littlejohn (1979a).

Dimensions in pillar-and-stall mines can vary considerably, and the areal ratio of pillar to stall is determined by the extraction rate. Extraction of 75% is

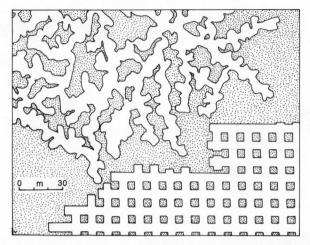

Figure 6.1 Plan of part of a gypsum mine in England, where about two metres thickness of gypsum has been partially extracted from a single, nearly horizontal bed. The random pattern of hand working dates from around 1930, and contrasts with the systematic grid of mechanized working from about 1960. The modern pillar-and-stall pattern has a 75% extraction rate, with no roof support except that provided by the gypsum pillars.

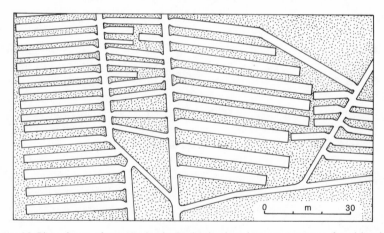

Figure 6.2 Plan of part of a coal mine in England, where just over a metre of coal has been partially extracted from a gently dipping seam. This is an example of hand working on a bord-and-pillar system, and timber props are placed for roof support as the bords are worked out.

a common optimum target, creating equal width pillars and stalls in regular stoop-and-room patterns. Old English coal mines often have this pattern with pillars about 2.5 m across. Many Scottish coal mines had stalls 5–8 m wide between elongate posts around 1–2 m wide, to give 80% extraction. Higher extraction ratios induced premature roof collapse, but were achieved in hand-worked mines where pack roof supports were built of dry-stone waste. Most coal mines have an extracted seam thickness between 1 and 3 m.

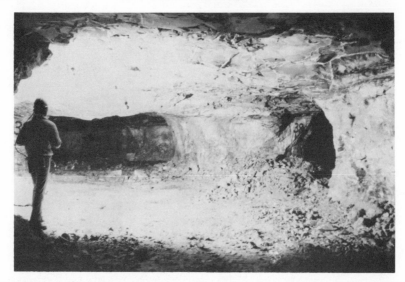

Figure 6.3 Pillar-and-stall working in a modern gypsum mine in Britain, with 75% extraction, stable pillars and stable roof spans of dry mudstone. The debris on the floor is dumped waste, not fallen roof. Just over 2 m of gypsum is extracted, and the mine lies 80 m below the surface.

Many stone mines in material stronger than coal could be worked with much larger galleries. Limestone mines are commonly stable with stalls over 10 m wide between slightly narrower pillars to give 80% extraction, in workings which may be 5–10 m high. Rocks such as gypsum, salt and soft sandstone demand lower extraction rates, as do very deep mines with greater overburden loads, but it is important to note that many old mines had their pillars trimmed, or robbed, until they were left in a critically marginal state.

The extent of mining should not be underestimated. The USA has 160 000 km² underlain by mineable coal, and already over 8000 km² are affected by subsidence. Britain has over 100 000 old coal mines and at least as many again worked for other ores. Most are small, abandoned and shallow and pose the maximum surface hazard. Many parts of Britain are riddled with old mine workings, but there is some recognizable geological control over the hazard.

As anywhere in the world, Coal Measure rocks are the most heavily mined, and only since around 1900, in both Britain and America, have more predictable total-extraction methods been applied. Not only coal has been mined, but also oil-shale, gannister, fireclay, brick clay, limestone, ironstone, sandstone and flagstone. Workings of all these can be very extensive; cavities up to 9 m high, beneath a cover of only 10 m of rock and drift, were found in 44 out of 100 boreholes on a site investigation in Leeds (Godwin, 1984), and Evans and Hawkins (1985) describe a fine example of old mines below a hospital site in Wales. The statistics of Telford new town in Shropshire (Whitcut, 1981) make sobering reading: 20% of the 7792 ha site is underlain by old shallow pillar-and-stall mines in 17 coal seams, 10 ironstones, 10 fireclays and also

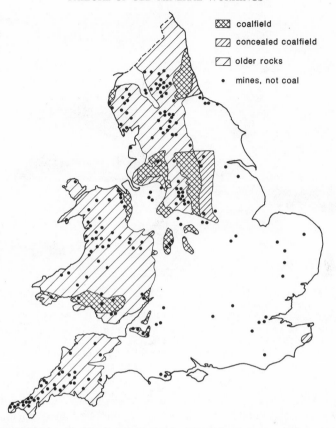

Figure 6.4 Distribution of old mines in England and Wales. Each dot represents a group of underground workings for minerals other than the Coal Measure rocks; opencast mines and quarries are not included. The coalfields have the highest densities of old mines, for coal, ironstone and other rocks. The concealed coalfields have few old shafts or workings into the coal beneath the thick cover of younger rocks, but are major areas of current mining. Older rocks are those from the Carboniferous and earlier periods.

brick clay and limestone, and there are also 2957 known mine shafts.

Rocks older than the Coal Measures include various igneous and metamorphic terrains which tend to be richer in mineral veins. These occur mainly in the west of Britain where there are thousands of old mines for iron, zinc, lead, copper, tin, tungsten, silver and gold as well as calcite, barite and fluorspar, with notable concentrations in the Pennine limestones and Cornwall (Figure 6.4). In contrast, the younger rocks of southeast Britain have only a scatter of mines, nearly all of which are pillar-and-stall workings in bedded mineral deposits. Old mines in the Jurassic and Cretaceous ironstones pose minimal subsidence hazard, and there is almost no subsidence over the vast areas of modern mining in the gypsum, anhydrite and salt of the Permo-Trias red beds. Old stone mines in the Cotswold and Bath oolitic limestones had up

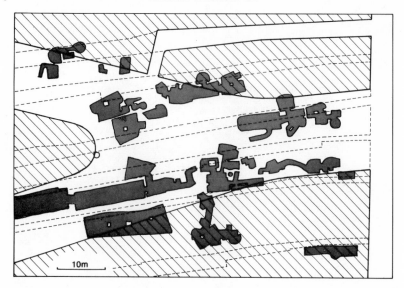

Figure 6.5 Mapped sandstone caves beneath buildings and streets in part of the city centre of Nottingham. The caves are nearly all cellars which were cut in the bedrock beneath old houses, factories and especially inns. Their distribution therefore relates to the narrow old alleyways (broken lines on the map) which led down to the market square (the open area on the right). Most of the caves were filled after the old buildings were demolished to make way for the new wide roads. There are undoubtedly more caves, open or filled, which remain unrecorded within this area.

to 85% extraction (Thomas, 1987). Chalk mines up to 8 m high and wide, including the Chiselhurst 'Caves', and smaller, older flint mines are hazardous where close to rockhead. There are clay mines in the Dorset area, and old sandstone mines are a hazard at various locations: parts of Castleford and Pontefract, in Yorkshire, are underlain by unstable mines in friable sand with 75% extraction beneath a cover locally of only 2 m of rock (Baldwin and Newton, 1987), and the city centre of Nottingham is underlain by a warren of cavities (locally known as caves) which are mostly old breweries, tanneries and store-rooms carved in the bedrock (Figure 6.5).

The subsidence mechanisms and their potential hazards over different mines relate closely to the geology. Vertical vein mines, in strong country rock, normally influence only a narrow zone of ground above them, and their great depths may preclude subsequent redevelopment of that land. Steeply dipping mine workings may effectively sterilize a wider zone of land on the down-dip side of the outcrop only, as over the old shallow gold mines in Johannesburg (Stacey, 1986); the dumping of mine waste is a common, appropriate practice in these zones. Mine workings in low-dip beds, notably coal, provide the most extensive subsidence. Figure 6.7 shows how this may include localized ground collapses due to roof failure (crown holes), subsidence over wider areas due to pillar failure, or less severe areal subsidence due to slow closure of the mine

Figure 6.6 Local subsidence and ground failure along the edge of a deep open-cut mine on a fluorspar vein in the English Pennines. The wallrock limestone is strong but here it has failed along an oblique fracture, and the large block in the photograph subsided and rotated until it leaned against the wall on the left.

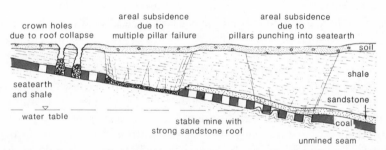

Figure 6.7 Profile to show main types of subsidence due to ground failure over old mines. The less severe area subsidence due to floor lift and pillar punching is not restricted to sites over flooded mines, but is more common where there is a saturated, weak floor rock such as seatearth or shale.

openings under greater overburden loads. The scale of subsidence over old mines can be spectacular; the town of Scranton, in Pennsylvania, provided an excellent example (Anon, 1916) when its main street developed a switchback profile with an amplitude of up to 3 m due to irregular extraction and pillar failure from multiple seams at depths as little as 20 m.

Pillar failure and areal subsidence

A natural restriction is imposed on pillar-and-stall coal mining by the low compressive strength, commonly less than 10 MPa, of in-situ fractured coal;

the constraint is on the percentage of a seam which can be extracted to leave pillars which ensure both short-term and long-term stability. Methods of calculating pillar strength are assessed by Hustrulid (1976), Bell (1978) and Sheorey *et al.* (1987), but in practice it may be difficult to quantify some of the parameters. The fundamental factors are the mass strength of the coal and the height–width ratio of the pillars, and these must be related to the extraction ratio and the overburden load.

Hucka *et al.* (1986) determine pillar strength S in a Utah coal mine as $S = C(0.65 + 0.35 w/h)$ MPa where C is the compressive strength of coal (locally 11.25 MPa), w is pillar width (13.7 m) and h is pillar height (5.2 m), to derive a pillar strength of 17.77 MPa. The pillar load L may then be calculated as $L = HD/(1 - e)$ where H is overburden depth, D is overburden density (commonly 22.5 kN/m^3 for coal measure rocks), and e is extraction ratio (0.5 in this case). At a depth of 120 m, $L = 5.4$ MPa, to give a safety factor of 3.3, though at 300 m depth the safety factor declines to 1.3. It should be noted that while this example demonstrates the principles involved, it does refer to a rather strong coal, worked with large pillars and an extraction ratio of only 0.5.

Mine pillars normally require a safety factor of at least 2 to offer permanent support. The usual causes of failure are the reduction of pillar width due to spalling, slaking or erosion (well illustrated by Evrard, 1987), or the loss of pillar strength due to weathering, oxidation or fracturing, besides any increased pillar load due to either structural surcharge or decline of the water table. Weathering and floor heave may be faster below the water table, but old mines above water suffer the additional hazard of fires. These are commonly started by spontaneous ignition where coal dust, air and water are present in the critical ratios (Dunrud, 1984). They are often the consequence of subsidence fissures opening airways to the coal, and then cause further severe subsidence as the pillars burn. Through 1986 a fire beneath the village of Oakthorpe, in the Leicestershire coalfield, caused severe problems, and some houses were demolished; though the fire was controlled by extensive grouting of the shallow old mines, the long-term effects have yet to be fully realized.

Pillar size is also critical to stability, and Orchard (1964) suggests that the minimum pillar width should be at least 10% of the overburden depth. He cites the Billingley colliery where severe subsidence occurred after only 40% of the coal was extracted but pillar widths were only 4–7% of the mine depth. When a pillar does fail, the roof load may be cast on to its neighbour and a domino effect may follow to create widespread ground movement. The classic case of failure was at the Coalbrook Colliery in South Africa in 1960 (Bryan and Bryan, 1964). A 300 ha section of mine collapsed within about five minutes, and as a result 437 miners died. It was a continuous failure of the pillars with no known initial trigger event. Mean depth of the mine was 140 m and 3–4 m of coal was being mined at 55% extraction, taking 6 m wide roads between square pillars which were 12 m wide but only 8.5% of the depth. Another smaller multiple-pillar failure in a West Virginia coal mine (Khair and Peng, 1985) may have been considered inevitable as the high extraction ratio left a pillar load well in excess of the tested strengths of coal.

Often known as troughs or sags, the areal subsidences due to pillar collapse are most common where mine overburden is between 15 and 30 m (Gray and Bruhn, 1984), and though size is related to the extent of pillar failure, most are less than 100 m across and a metre deep. Wider collapses are often contained by stronger roadway pillars, and though a gentle bowl is the normal surface expression, DuMontelle and Bauer (1983) describe marginal tension cracks and central pressure ridges. An area of 0.75 ha subsided by over 300 mm under the town of Bathgate in Scotland, and created ground strain of up to 7 mm/m. The movement occurred in three phases in 1975–77, significantly all in the winter, and in each case 30% of the subsidence was immediate, with the rest occurring through the following year; this is an instructive case history (Carter et al., 1981) of collapses in coal mines at depths from 12 to 34 m with a 75% extraction rate.

Pillar failures and surface subsidence over mines in other than coal are less common, due to the greater strength of most other rocks. Many old limestone mines in the Cotswold oolites have 85% extraction with no sign of failure. On the other hand, part of a working limestone mine in the Pennines failed in 1975, when a domino collapse of pillars 12 m wide and 5.2 m high under an overburden depth of 40 m caused immediate subsidence of a hectare of farmland above. A different mechanism of failure caused subsidence over old limestone mines at Wednesbury in the English Midlands in 1978 (Braithwaite and Cole, 1986). Collapse of the mine over an area of 90 × 140 m created a surface subsidence bowl 200 × 300 m, up to 1.2 m deep over a period of two months. The limestone pillars, even at the great depth of 145 m, did not fail, but a few metres of shale roof first failed across the stall spars and then crushed over the pillars of limestone; strength of the shale was between 39 and 50 MPa, roughly half that of the limestone.

Pillars may fail in shallow old workings due to newly imposed structural loads, though this is a decreasing hazard in mines at greater depths. Taylor (1975) shows that surface loading of around 300 kPa may add a structural surcharge, on pillars at a depth of 10 m, which is equivalent to 50% of the existing overburden stress. But at 30 m depth, the structural load is only 5% of the overburden load.

More commonly mine pillars deteriorate over time and eventually fail purely under overburden load. Ultimate closure of the mined voids occurs as a result of pillar crushing, roof overbreak, floor heave and pillar punching. With a compressive strength of 5 MPa for fractured coal, an extraction rate of 75% promotes inevitable pillar crushing beneath an overburden 50 m deep. Depths ranging from 45 m to 60 m are quoted by Cameron (1956) and others, beyond which old coal mines are rarely found open. The pillar crushing does cause surface subsidence, and is most marked where strong sandstones form the floor and roof of the mined seam, but it may be ameliorated by other processes.

Roof overbreak, with its consequent bulking of debris, may aid closure of the mined void without surface movement, and also offers lateral support of the pillars; but a consequence may be pillar crushing in weak roof rocks between migrated roof collapse domes. Floor heave, due to expansion of the exposed floor rock, is of limited extent; it is more pronounced in some clay seatearths than in other rocks, is more when the rocks are wet, and ultimately offers similar lateral support and void closure. Alternatively pillar punching occurs where pillars sink into a deformable floor rock, and consequently promote surface subsidence. Piggott and Eynon (1978) suggest that punching will occur wherever the pillar load exceeds 3.7 times the strength of the floor rock, which means that is likely to be common where mines with 75% extraction have a wet clay floor and lie at a depth of over 40 m. Punching is known at depths of only 15 m. The surface subsidence tends to be a slow creep often lasting for over a year; it is a common cause of shallow areal sag subsidences in the USA (Bruhn et al., 1981), and Orchard (1964) described a subsidence bowl 600 mm deep over a punched coal mine in Northumbria.

The inevitable long-term collapse of old, over-extracted coal mines means that the subsidence hazard does eventually decline, though it may last for over a hundred years in some cases. Stephenson and Aughenbaugh (1978) describe ground conditions where areal subsidences had occurred over collapsed mines which had been abandoned for 50 years. Boreholes found numerous small voids, averaging only 110 mm high, at all depths from the surface to the old mines 90 m down; upward void migration and roof bed separation was clearly continuing and there was still a risk of further surface subsidence. A Pittsburgh coal mine, closed in the 1920s, promoted areal subsidences in 1964, 1972 and 1973 all on almost the same hectare plot of ground (Bruhn et al., 1981). One time prediction that is possible is that a migrating subsidence wave from longwall undermining may commonly cause pillar failure and accentuate ground movement over even very old mines.

Steep dips are a complicating factor in any mine pillar failures, and old workings dipping more than 15° require particular attention, especially where there is a strong sandstone roof. The extensive ground failures at Bathgate (Carter et al., 1981) were over 200-year-old mines dipping at 22°. Stacey (1986) describes the additional hazards from cantilever action of the hanging wall over old gold mines at even steeper dips beneath Johannesburg.

Crown holes

Crown holes are localized surface failures created where progressive roof collapse causes a void to migrate upwards from an old mine working and eventually break the surface (Figure 6.8). They are equivalent to subsidence sinkholes in limestone areas, and are also known as plump holes, cave-ins, sinkholes, pit subsidences, pits and sits in different mining regions.

At the surface, crown holes are rarely more than 5 m across or deep, and

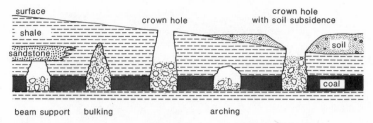

Figure 6.8 Styles of roof failure in abandoned coal mines leading to either ultimate roof stability or crown hole development in the surface.

most commonly have around half those dimensions. As they are the direct result of roof collapse between stable pillars, they mostly form over the wider stalls of shallow old mines; Dunrud and Osterwald (1978) describe a spectacular Wyoming example of numerous crown holes almost matching the pattern of the mine 15 m below. Thick drift may flow into a migrating void when it reaches rockhead, and may then create a broad saucer-shape subsidence bowl (Figure 6.8), instead of a conventional small crown hole, and where a number of stall roofs fail the surface subsidence may coalesce into an even wider bowl. Thorburn and Reid (1978) describe a case in Lanark, Scotland, where some houses had to be demolished due to severe subsidence; excavation revealed that pillars in the old mines 15 m below had not failed but there was extensive fracturing and sag of the rock and drift cover, with incipient crown hole development and up to 338 mm of surface lowering.

Crown hole development is strongly influenced by the cover rocks between the mine and the surface. Void migration, or upward stoping, may be stopped by stable arch formation, by beam action or by bulking which eliminates the void. Arches are formed in all rocks, and field measurements of old mines revealed in later quarry faces indicate that the collapse height is normally only 2.68 times the span of the mine working (Garrard and Taylor, 1987), with flatter arches forming in stronger sandstones. Piggott and Eynon (1978) suggest that a bed free from joints and of thickness greater than 1.75 times the roof span will prevent void migration. Crown holes are limited by any competent rock, but conversely are known from depths of 50 m below thick soils (Gray and Bruhn 1984).

The second depth restriction is from the bulking factor which causes the fallen debris to occupy a larger volume and eventually support the roof. The vertical extent of collapse before bulking causes closure is a function of the void shape (whether it is parallel-sided or conical), the mined-out thickness, and the bulking factor (ranging from about 25% in some shales to 50% in stronger rocks). Theoretically the limit is therefore 2–12 times the mined thickness, with the worst case presented by conical collapse in weak shale (for full graphical treatment, see Piggott and Eynon, 1978, or Dunrud, 1984). However, sloping mine floors may allow debris to run or be washed away and hence allow migration to proceed further. Similarly, steam weakening above

Figure 6.9 A crown hole which opened up partly beneath a heavy duty quarry access road in Lincolnshire. Old pillar-and-stall ironstone mines underlie the site at a depth of about 25 m.

a burning coal mine has caused migration over 15 times seam thickness (Dunrud, 1984). Overall, crown holes are rare above mines at depths greater than 30 m, though hazard zoning (see below) may demand a more conservative view.

The surface failure at a crown hole may be instantaneous, but this will follow many years of hidden void migration. Prediction of events is impossible and some crown holes have formed over mines hundreds of years old. However, shallow movements are related to drainage and many crown holes form during or soon after spells of wet weather; Gray and Bruhn (1984) observed a three to eight month time lag between high-rainfall events and crown holes in the Pittsburgh coalfield, but Nishida *et al.* (1986) describe a more direct and much shorter correlation in the coalfields of Japan. Also in Japan, the same authors describe a major increase in crown hole development promoted by an earthquake in 1978 (with peak vertical acceleration close to 0.1 g); almost any abnormal ground disturbance, whether natural or due to engineering works, is liable to induce or accelerate crown hole collapses.

Abandoned mine shafts

Old mine shafts present a localized but potentially very dangerous surface failure hazard, especially where they are inadequately covered and forgotten. Some shafts, over 30 m deep, predate 1600, but most originate from between 1700 and 1900. Before 1800 most shafts were less than 200 m deep and from 1 to 4 m in diameter; those in hard-rock metal mining areas were generally

Figure 6.10 Collapse of an old mine shaft at Bishopbriggs, Glasgow. The site was the garden of a house, but the shaft had only ceased operation 16 years previously, before being inadequately covered and forgotten (Photo: Press Association).

narrower than those in coal measures. More recent shafts tend to be larger (see graphs of changes in Littlejohn, 1979a), and by 1900 were often 500–1100 m deep and 2–7 m in diameter. Shafts were commonly paired and very close together; only since 1887 have mining regulations, in Britain, ensured a minimum separation of 13.6 m.

In some mining regions there are enormous numbers of abandoned shafts. In Britain, various estimates all claim over 100 000 old shafts in the coalfields and many more than that number in the old metal mining fields (Figure 6.4). On one 7 km stretch of motorway route near Birmingham, 55 shafts were traced from old records and another 19 were found during construction (Morse, 1967), but a new road across parts of the Derbyshire lead mining field could encounter far greater numbers. In addition, the coalfields have similar numbers of adits or sub-horizontal mine entrances, driven from outcrop, and also lost, buried or collapsed.

Long after their abandonment, shaft locations may or may not be known, and their condition and stability may vary enormously. Even early attempts to render shafts safe were often inadequate, and failure of shafts more than 50 years old is all too common. Some were left open when abandoned, while others gained protective walls or domes, but the major threat is from those poorly capped or filled which may now have no surface indication of their presence; Whitcut (1981) gives an impressive list of the variable factors found in old shafts in Telford new town. Many shafts were capped with timber, a single steel plate, or perhaps old railway lines, commonly with poor founding for the cap. Others were partly filled with ash or mine waste, which may rest on a timber platform, or staging, at or below rockhead, or even on a tree jammed in the shaft wherever it caught; even if the fill was to the bottom of the shaft, roadways left open allowed flow and settlement (National Coal Board, 1982). Also the shaft lining, of timber, brick, stone or concrete, depending on site and age, may be in a parlous state, especially above rockhead.

The main mechanisms of subsequent shaft collapse are the failure of timber staging below fill, the failure of fill where roadways are not sealed off, lining failure above rockhead, and collapse of inadequate cappings. Dean (1967) and Gregory (1982) give instructive descriptions of a variety of shaft failures; both refer to the famous and dramatic loss of a complete train when a colliery shaft failed beneath a railway yard near Wigan only 14 years after it was abandoned and inadequately filled. Morse (1967) describes a borehole revealing 20% voids down a supposedly filled shaft 161 m deep, and Cunningham and Sutherland (1984) provide a useful description of subsidence damage to a house due to compaction of a shaft fill. An open lead mine shaft was recently found, fortunately before it collapsed, immediately beneath the slabs of a pavement being rebuilt in a Derbyshire village; whereas in 1977 a coal mine shaft did fail, breaching a trunk road near Birmingham. The latter had a concrete cap, placed before the road was built 53 years earlier, but the shaft shoulders had decayed and the minimal cap slab had dropped into it. Prediction of the timing and location of old shaft failures is next to impossible.

Drift and soil thickness at a shaft site also may be important. Collapse into only a narrow shaft in rock may allow formation of a wide surface crater in thick drift; Gregory (1982) describes such subsidence craters up to 60 m in diameter.

Bell-pits

A medieval method of mining created bell-pits where a very shallow seam of coal, or ironstone, was most easily worked by way of numerous shafts, so saving on underground transport. The shafts were mostly around a metre in diameter, could be up to 12 m deep, and widened to a room in the coal which was cut to

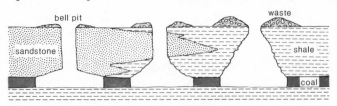

Figure 6.11 Diagrammatic profile through a series of bell-pits into a coal seam, with various shapes of pit related to the overlying rock.

8–20 m across before being abandoned at signs of imminent collapse (Figure 6.11). The next bell-pit was then sunk nearby, and densities of over 200 per hectare are not unknown. Though bell-pits may literally perforate the ground, and offer a distinctive subsidence hazard, they occur at only very shallow depths and few post-date 1700.

Site investigation

Old mines provide an obvious threat to civil engineering works, but site investigations commonly encounter severe problems due to lack of recorded information. In Britain and other areas with a similar long history of industry and development, it is reasonable to assume that any coal seam or mineral vein has been worked. Lack of data is no evidence of lack of mining; this is demonstrated in the Wigan area, where 1200 shafts were recorded in 1958, but the same area held 1700 known shafts by 1980, and the real total remains in the realm of guesswork.

The desk study search of available data assumes a major importance where there is any possibility of past mining. The sources of information vary between sites, but Table 6.1 lists the major useful sources in Britain. British Coal (formerly the National Coal Board) has a legal obligation to record and provide data on all coal mines, but their records of older mines cannot hope to be complete. Outside the coalfields, mining records are generally less well organized on a county basis, whose local authority is therefore the starting point for a search. The Geological Survey has its own records, and old manuscript maps, both geological and topographic, are commonly invaluable (Charman and Cooper, 1987). Other sources include air photographs, which may indicate old shafts, bell-pits and some shallow, partly-collapsed workings (Giles, 1987), together with local consultants and specialists, and various archive collections. Symons (1985) gives good outlines of desk study procedure in British coalfields, with both exhaustive and more practical and economical versions.

It is significant that, in a subsidence assessment in South Wales (Statham et al., 1987), the British Coal and Geological Survey records yielded 83% of the

Table 6.1 Sources of data on past mining in Britain.

1. In coalfields: British Coal Area Office for:-
   subsidence report on current and future mining
   seam maps of current and recent mining
   abandonment plans for seams no longer worked
   register of shafts and mine entries
   records of conditions, capping and filling of old shafts.

2. Outside coalfields: mining records with either the Local Authority,
   the Mineral Valuer, or Health and Safety Executive (Bootle).

3. a) Published 1: 10 000 geological maps
   b) British Geological Survey for old maps, records and unpublished data
   c) Old topographical maps.

desk study data for 30% of the effort. Old maps, local authorities and local consultants yielded another 12% of data for 12% of the effort, and other sources had minimal yields. Furthermore, Carter *et al.* (1981) describe their desk study at Bathgate, Scotland, which provided positive data on mining in less than 30% of the area which was subsequently found to be undermined. Any complementary site visit may or may not find visible signs of old workings, shaft tops, old tramways or spoil heaps.

Field investigation, on a thorough and expensive scale, is therefore frequently unavoidable; but, as site investigation is always only a small fraction of total project costs, excessive economies are misguided and false. Fieldwork separates into the two aspects of searching, for old workings at depth, and for isolated shafts.

Though trenches and trial pits have their value in searches for very shallow features, old workings are normally located by drilling, when penetration rates, fluid loss, and the RQD on cored holes are the most useful features to watch for. Previous desk study should suggest the likely depth of any workings, and depth and spacing of boreholes can be planned from this and then modified as work progresses (see for example Taylor, 1975). Healy and Head (1984) suggest holes should reach a depth at least 10 times the likely seam thickness or mine height, but this is a bare minimum, and Littlejohn (1979b) suggests that at least on major sites some key holes should reach 60 m to establish strata control. Drilling to only 15 or 20 m is common but bad practice, while Taylor (1975) has seldom found the need for holes deeper than 45 m; in coal measures, 30 m may be an appropriate depth guideline, unless the local geology suggests otherwise. On the other hand, experience in Nottingham, where small irregular cavities are found in the strong sandstone (Figure 6.5), has found it adequate to drill just 5 m below formation or pile footing and then test the hole with high water pressure; in some cases splayed test holes are also drilled on critical sites.

Spacing of exploration holes will depend on site conditions, but should never be on a regular grid in case this coincides with a regular pattern of mine pillars; and this coincidence has happened! An 8 m grid of holes on a site in Kent completely missed a series of very shallow chalk mines (Winfield, 1984), which were only revealed during subsequent pile driving. An irregular spacing as close as 5 m or less may be appropriate beneath foundation zones on some sites; in addition, exploratory holes should be placed outside the perimeter to a distance of one third of the likely depth, due to the potential angle of draw from adjacent mine failure (Littlejohn, 1979b).

To assess the size of encountered cavities, downhole television may be used. However, Braithwaite and Cole (1986) found this had a range of only a few metres in the large flooded limestone mines below the West Midlands, and image intensifiers and computer enhancement will be needed for useful results. But in the same mines, ultrasonic scanners were found to be successful and had a range of up to 50 m (Cook, 1984). Surface geophysical surveys are of almost no use in this field, as the various techniques either have depth limitations of around 5 m or can detect voids at a depth of only twice their diameter.

Searches for old shafts buried at only shallow depths may, however, usefully employ a variety of geophysical surveys, though they may be limited by background interference on urban sites. Useful reviews are provided by Mason (1984), Gallagher et al. (1978), McCann et al. (1987) and Bell (1986). The magnetic technique is the most useful, involving a walking-pace survey with hand-held proton magnetometer, to identify anomalies which are much wider than the shaft and often of a clearly recognizable dipole pattern (McCann et al., 1987). Used on a motorway construction site, a magnetic survey found 7 out of 17 shafts and bell-pits (Burton and Maton, 1975). On a test site in open country (Gallagher et al., 1978), the magnetometer found all seven shaft sites, though the lack of dipole anomalies at most suggest that it was identifying shaft-head structures and not the shaft itself.

Electrical surveys, measuring ground resistivity, are dismissed by most authors with respect to shaft searches as being ambiguous, impossible to interpret, or having an unacceptably low recognition rate. The inductive electromagnetic technique is a simple and faster improvement on the resistivity survey, but still has a poor record of achievement in shaft detection. Ground probing radar (reviewed by Darracott and Lake, 1981) suffers from a depth penetration of only a few metres, especially in wet clays; Leggo and Leech (1983) found that a trolley-mounted antenna could recognize a shaft directly beneath it, as long as any cap or plug was less than a few metres thick, but Sladen et al. (1984) could achieve no satisfactory correlation between radar characteristics and known mine workings at depths of only a few metres below a site in Canada. Gravity and microgravity surveys are of no value as background noise masks the very small anomaly of a shaft, and seismic surveys fail as the target is too small. A geochemical resource is offered by sensitive methane detectors, and anomalies of 10–100 ppm in soil or open air can indicate coal mine workings or a shaft (Healy and Head, 1984).

Drilling or trenching to locate buried shafts both suffer from the hazard of collapse during the operation. Safety techniques include mounting the drill-rig on long girders (Healy and Head, 1984), using an arm-mounted drill (Littlejohn, 1979b) or trenching with a dragline (Maxwell, 1975), and the crew should also be on safety lines. Boreholes need to be on a grid with spacing less than the expected shaft diameter, either drilled systematically or starting at the most likely site and then following a sequence in an outward spiral. Trenching is cheaper, though limited in depth potential, and may expose remains of brick linings or some alien fill within disturbed drift.

The costs of shaft searches were reviewed by Mason (1984). Magnetic geophysics costs £350 for a half hectare search, and trenching to just 3 m depth costs £500 for the same area. Bored probes cost up to £600 if the shaft is found within 5 m of its suspected position, but wider searches cost £25 × D^2, where D is the distance in metres between true and suspected locations.

These figures point to the value of careful desk study and any geophysical indications.

Treatment of old workings

Construction in areas of extensive old mine workings normally requires some form of ground treatment or remedial action. The main options are excavation and backfill, consolidation with grout, piling to below the mines, raft foundations for buildings, or relocation of structures (Figure 6.12). Individual large structures may be most economically resited on stable ground, or a variation on this is to move the building to an area of shallower mining where ground treatment is then less expensive (as in Evans and Hawkins, 1985). Land values may determine the feasibility of treatment; limestone mines beneath Telford are relatively stable compared with some old coal mines, but are so large than grouting is uneconomical and development over them is therefore avoided (Whitcut, 1981). Gray and Bruhn (1984) claim that large-scale treatment of old coal mines is uneconomic for residential development in areas in the United States, though others present a different viewpoint for projects in Britain (Higginbottom, 1984; Slowikowski, 1978).

There is rarely call to grout old mine workings where there is a cover of more than about 25 m of competent rock, though local site conditions may determine otherwise. Failures of old mines are less common at greater depths, and any surface subsidence effects are also normally reduced, so that appropriate foundation design may be an adequate precaution. Though the total loads on mine pillars increase at depth, the added structural loads are a lesser proportion; in marginal cases the pillar loading can be calculated to assess the need for grouting (as in Corbett, 1984). The literature on old mine consolidation is extensive, with good reviews presented by Littlejohn (1979*b*), Littlejohn and Head (1984), Hislam (1984), and Healy and Head (1984). Useful case studies describe ground treatment below road projects (Brown and Buist,

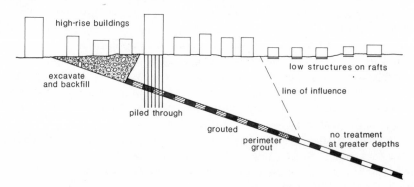

Figure 6.12 Alternative methods of ground treatment and foundation design appropriate for construction over abandoned pillar-and-stall mines.

1976; Ferguson, 1986) and for major building development (Price *et al.*, 1969; Forrest and Anderson, 1984; Evans and Hawkins, 1985), while Thomas (1987) compares the cost of alternative remedial works for construction over some shallow old limestone mines.

Grouting of old mines is normally carried out with a mix of cement, sand and PFA in proportions ranging around 1:1:5–20, injected through boreholes of 50–75 mm diameter. These are drilled on a grid, which may need to be irregular, with spacing between 4 and 6 m, and a typical take is then between 3 and 12 tons per hole. The take will vary with the extent of collapse in the mine and also the amount of stowed or packed waste, and considerable local variation must be expected; also the first hole into a new area of mainly open mine may take well over 100 tonnes. Where the take is high, secondary holes are best employed on half the grid spacing. Grout pressures should not exceed 10 kPa per metre depth, to avoid uplift, and holes should be cased through any drift.

The area to be grouted should underlie all planned structures and also should extend through a marginal zone of width 75% of the depth to allow for draw. Also the main infilling can only follow placement of a perimeter grout zone which is normally injected through a line of holes only 1.5 m apart. Horizontal spread of the perimeter grout may need to be controlled, especially in more open mine workings, and it is usual to modify the grout mix with the addition of bentonite to equal the cement, and also variable quantities of sand or pea-gravel. Further details on grout specifications may be found in Hislam (1984) and Littlejohn and Head (1984), and the efficiency of the grouting may be assessed by checking that permeability is less than 10^{-5} m/s on a packer test.

Old mines with minimal collapse and largely open workings may be more economically filled with inert material dropped down larger diameter holes, purely under gravity or hydraulically stowed if free drainage is available. A rock paste, largely of old colliery spoil, has been successfully used in old limestone mines (Cole *et al.*, 1984) and has been found to flow up to 100 m from the feed hole (Cole, 1987*b*). Inclined mines are more easily filled in this way. Steeply-dipping gold mines below Johannesburg were treated by gaining access and then forming elongate pillars of mass concrete extending down the dip; these were 2 m wide, on 10 m centres, and voids between were backfilled with loose waste above a barrier pillar formed along the strike (De Beer, 1986). A comparable technique involves partial grouting where a strong roof is present, creating pillars below drilled holes, either as conical heaps or batch-placed pancake pillars, of viscous, pea-gravel grout.

Excavation and backfilling of old mine workings is normally only feasible to depths of around 5 m, though a site in Leeds was excavated to a depth of over 9 m, through boulder clay and coal measure rocks which did not require blasting (Winfield, 1984). A benefit of this technique is that the extracted coal is saleable, and any planned basements to a structure can economize on backfill.

On the other hand drainage problems may be prohibitive below the water table, and urban sites may lack the space for a large muck-shift.

Stabilization by promoting collapse, perhaps by pillar blasting, is rarely applicable, though Stacey (1986) describes dynamic compaction to collapse a cantilevered hanging wall and effectively close the voids in shallow, steeply dipping mine workings under Johannesburg. The rockhead outcrops of the same mines were plugged and capped with concrete beams, but this is only feasible where dips are steep.

Foundation piles taken below unstabilized mine workings (Healy and Head, 1984) have only limited application even when of large diameter and sleeved. They are normally limited to depths of less than 30 m, and where they can be driven through drift or shale, as pre-boring through sandstone is rarely economical; further old mines at greater depths, any likelihood of future longwall undermining, or steep dips, where oblique collapse may impose shear loads, all preclude their use. Where potential subsidence is less, the alternative to ground treatment is appropriate foundation design—either rigid rafts, to bridge local failures and spread ground loadings, or flexible structures where the old mines are not so shallow as to threaten excessively

Figure 6.13 Old sandstone caves in Nottingham preserved beneath the Broadmarsh shopping centre. The caves were artificially cut into a sandstone cliff, and much later, some nineteenth century brick foundation columns were built through them. The cellular concrete roof in the photograph, which is the shopping centre floor, spans the caves and is supported to the left on a combination of solid rock and some concrete piers which had to breach the caves.

high ground strain. The literature on design techniques is extensive (see Healy and Head, 1984, and various papers in Forde *et al.*, 1984).

Total costs of ground treatment in areas of old mine workings are estimated at usually 3–5% of housing costs (Higginbottom and Head, 1984). Slowikowski (1978) similarly found ground treatment (including grouting, regrading and some raft foundations, but without meeting any potentially expensive shafts) costing an average of £495 per house on a site in Wales; this he considered worthwhile as it still saved £400 per house on the costs of good green land, and was also a socially desirable utilization of derelict land. In Telford new town, costs of ground treatment in 1980 were £10 000–£30 000 per hectare (Whitcut, 1981)—again at around 3–4% of the total development budget, but Evans and Hawkins (1985) describe grouting costs of £880 000 for a 5 ha hospital site on difficult ground with extensive undermining.

Treatment of old mine shafts

Shaft treatment is an expensive operation for a small area of ground, so relocation of structures may be worthwhile, and then a less substantial cap may be appropriate for a shaft left in undeveloped ground. The available methods of treatment are capping, filling and grouting, or a combination of these.

The basic feature of a shaft cap is a square reinforced concrete slab with edge length at least 2.5 times the shaft diameter and firmly founded at rockhead (Figure 6.14). An appropriate slab thickness is 450 mm, though less may be acceptable over small diameter shafts, and over those where development is clearly not planned and a monument is left to mark the position and prevent unsuitable covering. Variations on design are given by Healy and Head (1984) and National Coal Board (1982), and Dean (1967) illustrates a very massive cap for a shaft beneath heavy industrial plant. Shafts to be covered by roads or buildings should be filled as well as capped, and where filling cannot be reliable, perhaps due to obstructing staging within the shaft, a more massive cap is required; Morse (1967) describes such a shaft on a motorway line, where a

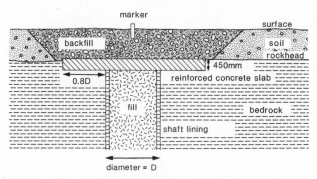

Figure 6.14 Appropriate design of a capping for an abandoned mine shaft where rockhead is not too deep.

grouted cap 12 m in diameter and 18 m deep was formed in the drift.

Shafts may be filled with any available hardcore, though Dean (1967) recommends that good, hard stone all sized between 150 and 400 mm should be used for the first 15 m above any side roadway to prevent flow into it. The fill may then be consolidated with a drop-ball, capped and if necessary grouted for the top few metres through holes left in the cap. Where roadways from the shaft are stopped off, drainage can be important, and Gregory (1982) describes the disastrous failure of fill, in a Lancashire shaft over 600 m deep, which liquefied when drainage was interrupted.

Old shafts with fills of unselected and uncompacted waste may require grouting to permit safe over-building. For shafts less than 2.5 m in diameter, one central grout hole is adequate, but larger shafts should have four holes; and all are drilled with a rig safely mounted on long girders. Morse (1967) describes a filled shaft 161 m deep which was grouted from the bottom up at 3 m intervals; the take at each stage was mostly a few tons or almost nil, but two points took over 200 tons each at the level of open roadways.

Current costs for a shaft treatment generally fall in the range of £1000–15 000. An added complication in Britain is that all old coal mine shafts belong to British Coal, so treatment can only proceed after due consultation and indemnification.

Subsidence hazard zoning in old mining areas

Surface hazard zoning with respect to old mines is fraught with problems largely due to the difficulty of establishing reliable parameters, but is often necessary to ensure economical land utilization. Over 90% of surface failure events are crown holes, and a wealth of experience, especially in coal measure rocks, allows some degree of hazard prediction, but there is considerable variation of opinion, and local information must always be regarded as significant.

Crown holes are generally limited by the thickness of roof cover to rockhead, and most holes occur where the cover is less than 30 m. However, Gray and Bruhn (1984) find most American cases in a cover of less than 15 m, while Healy and Head (1984) claim that in Britain they are rare only with a cover of over 70 m, and boreholes over old mines in Illinois found migrating voids up to 80 m above the workings (Stephenson and Aughenbaugh, 1978). Over old limestone mines in the English Midlands, 77% of crown holes were through cover of less than 30 m, but 3% were through over 70 m (Ove Arup and Partners, 1983; Cole et al., 1984). The limit may be alternatively expressed in terms of the migration ratio (H/T), defined as the thickness of rock cover H divided by the extracted thickness of mineral T, but various studies on crown holes cite maximum values of the migration ratio anywhere between 4 and 18. A major study on crown holes in the South Wales coalfield (Statham et al., 1987; Payne, 1986) found 90% of crown holes where H/T was less than 6, and

Figure 6.15 A sinkhole opened up in the floor of a mine tailings pond in the English Pennines, when it was being partially excavated by dragline. The sinkhole lies over an open vertical stope in an old lead mine; the deepest tailings, just out of sight in the photograph, rested on only a thin soil and timber beams across the old mine.

also within 100 m of the seam outcrop; but some crown holes were found where H/T was up to 8 in areas of low dip, and up to 18 where steep dips allowed fallen debris to run away. The same study found 76% of events were at mine entries or where cover material was all drift, and only 24% were through a rock cover. Distributions of subsidence sites with respect to cover depth over old mines vary with the styles of failure developed in the particular area. Figure 6.16 compares data from South Wales and Pittsburgh, USA, where the former area has many adit entry collapses (buried beneath very shallow soils), while the latter has more trough subsidences (formed over deeper mine pillar failures), and both have crown holes formed through the lower cover thicknesses.

With these variations in crown hole and subsidence parameters, any guidelines for hazard zoning can only be approximations, and over-simplified rules can be dangerous. A good example of zoning was an Edinburgh housing

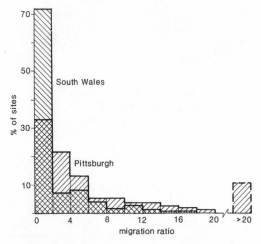

Figure 6.16 Thickness of cover at subsidence sites over old coal mines. Percentages of the total number of recorded subsidences are plotted against the migration ratio (cover thickness/mined thickness) for two areas. The South Wales data cover 309 subsidence sites including a large number of old adit entries (after Statham *et al.*, 1987) and the Pittsburgh data cover 239 cases including a large number of trough subsidences (after Bruhn *et al.*, 1981).

scheme planned to avoid all old mines less than 18 m deep and to limit structure sizes over any possible workings down to 30 m deep (Price *et al.*, 1969). Zoning over some mines in Canada restricted development where cover was less than 20 m in areas of low dip, and less than 30 m in higher dips, and also required raft foundations in an intermediate zone to twice these cover depths (Sladen *et al.*, 1984). Comparable results are achieved through delimiting a zone of restriction (of no development without ground treatment) by a migration ratio of 6, and an intermediate zone (of specific foundation requirements) by migration ratios of 6 and 10 (Statham *et al.*, 1987; Higginbottom, 1984); the latter paper also provides a good example of a more detailed zoning scheme, and makes the valid point that wider zones are wasteful of land resources. An example of zoning by the depth to probable old mines is incorporated in a generalized ground hazard map in Figure 13.2.

Exceptions to any guidelines may be imposed by various factors, and some noteworthy examples are quoted by Carter (1984). Hazardous zones should embrace thicker cover or higher migration ratios in both weaker rocks and any rocks at steeper dips, and migration ratios should treat the thickness of multiple seams as cumulative. There is no real safe age limit when old mines can be expected reliably to achieve stability, and hazards are always increased by any mine dewatering, excavation which decreases cover over a mine, or added drainage into a mine. On the other hand, very old mines may well not have been worked below the water table (Price, 1972), so reducing the area of potential hazard.

The major hazard of areal subsidences due to pillar failure is from old,

Figure 6.17 An abandoned limestone mine in Derbyshire. The roof is strong limestone, with only little block collapse; it is supported by a combination of in-situ rock pillars (on the right in the photograph) and well-constructed packs of waste stone (on the left and in the centre).

shallow mines, and these lie largely within zones already defined above with respect to crown hole failures. There is, however, no real depth limit beyond which mine failure can be ruled out; pillar collapse in coal mines is already known at depths of 140 m, though partial-extraction coal mining from depths of much more than 120 m is commonly not practicable because of the very low extraction ratios which are possible. Sensible hazard zoning is barely feasible, and it is fortunate that this type of subsidence hazard is a relatively rare event over deep mines. In Bathgate, Scotland, development was restricted in a zone reaching 60 m from the coal outcrop, so spanning mines to a depth of 25 m, and yet a domino-style pillar failure caused subsidence which reached 275 m from the outcrop (Carter *et al.*, 1981). Certainly site investigation boreholes should reach 60 m beneath any major structure where old mines may possibly exist, and borehole siting should allow for an angle of draw of failure between 15° and 20°. Attempts to establish monitoring programmes to predict this style of subsidence suffer from the need for an excessively large data base (DuMontelle and Bauer, 1983) and so have limited value.

The higher strengths normal in rocks other than coal measures permit deeper mining and, except under modern conservative mining regulations, therefore create larger potential hazard zones. Limestone mines in the English Midlands have recently caused ground subsidence due to failure at a depth of 145 m (Braithwaite and Cole, 1986); the previously-used hazard zone, based on a cover of less than 60 m, was rendered irrelevant, and a new wider limit is not practicable. In Johannesburg, no surface development is permitted over untreated old gold mines less than 90 m deep, and there are no restrictions at all only where the steeply dipping mines are deeper than 240 m (Stacey, 1986). This tends to sterilize a zone of valuable urban land up to 500 m wide down-dip of the outcrop, but as subsidence decreases with time and is normally minimal after 10 years, there may be a case for relaxing regulations at a suitable interval after mining is completed (De Beer, 1986).

Any schemes of subsidence hazard zoning must also incorporate any isolated shafts; adit entries normally fall into the established hazard zones of potential shallow workings. Restrictions may be applicable in a zone of diameter twice the thickness of drift centred on any proven shaft, but increased appropriately where there is any doubt concerning the precise shaft location. This reflects the critical value of old records and direct site investigation in establishing any hazard zoning scheme.

Backfilled quarries

A special subsidence hazard is provided by old quarries and surface mines which have been backfilled, and often forgotten. Modern quarries tend to be large, well-documented and restricted to few rock types, but old workings are much more numerous and in a huge variety of materials. Ground investigations at Telford new town site revealed old quarries in clay, marl, shale, sand, gravel, sandstone, limestone, dolerite, basalt, fireclay, coal and ironstone (Whitcut, 1981), and the historical demand for local building materials repeats this pattern widely.

Most old quarries were less than 20 m deep, but the fill material may have high potential compaction. Even a modern backfill in a continuously worked opencast coal mine may consist of cohesionless rock fragments with a porosity up to 40% and containing voids up to 0.5 m across (Charles et al., 1978). Very different are lagoon fills of fine-grained mineral-separation tailings, which can have high potential compaction and may liquefy when disturbed. Old random fills may be extremely variable in nature and properties, as summarized by Charles (1984). Domestic refuse offers the lowest stability; within a 10 year period it may settle by 10–20%, and the settlement is only in the lower end of this range when the material is compacted to a density of over 1 tonne/m^3 when laid.

Self-weight compaction, and consequent surface settlement, plots linearly against the logarithm of time (Charles, 1984); on modern backfills it is normally effectively complete within 3–5 years, but may be reactivated by structural loading. Total settlement is commonly around 1% of depth on deeper fills, which are partially compacted during accumulation, but may be a higher proportion at shallow sites. Settlement is also increased by a rising water level, commonly due to water-table rebound on cessation of pumping at a deep opencast mine; saturation collapse may be up to seven times normal compaction rates (Charles et al., 1978) and additional ground settlement is commonly around 5% of head increase. Various papers in Geddes (1985) describe settlement on modern backfilled sites, and Dunrud (1984) reviews similar experience in America.

The main hazard from backfilled quarries arises where they remain unrecognized during subsequent redevelopment, and there are all too many cases of building damage due to differential settlement across a buried quarry edge. Younger backfill sites may survive as derelict land, but older ones may have no surface indications. Old topographic maps are a valuable source of data where old quarries are suspected, but Browne et al. (1986) describe unrecorded quarries in Glasgow which were opened and backfilled between successive revisions of the ordnance survey. Redevelopment over any old quarry will normally require a trial pit to determine the nature of the fill, and the settlement may be measured by a field test with a loaded rubbish skip over a period of a month (Charles and Driscoll, 1981).

Small old quarries with questionable fills are best avoided during urban redevelopment, and open space requirements normally make this feasible; redesign of an Edinburgh housing development, described by Price et al. (1969), demonstrated this well by avoiding both an old gravel pit and an old filled limestone quarry. Modern backfilled opencast mines are commonly built over after a suitable delay, which may be less than ten years for low rise housing on raft foundations (Penman and Godwin, 1978), as long as excessive differential settlement across any buried quarry edge is avoided. Delay times may be shortened by use of preloading under surcharge, dynamic compaction or vibrocompaction (Charles, 1984); however, these techniques only reduce, and do not eliminate, further collapse settlement due to a rising water table (Charles et al., 1978), and preconstruction inundation may be a better way to reduce subsequent settlement.

7 Salt subsidence

Salt occurs as a rock most notable for its extremely high solubility in water. Consequently groundwater flow in contact with salt causes extensive solution, cavity formation, collapse and ground subsidence. Natural salt subsidence affects large areas over long time spans, but is greatly accelerated and usually localized by artificial brine extraction.

Salt (strictly the mineral halite, NaCl, also known as rock salt) occurs as a sedimentary rock precipitated by solar evaporation of lake or sea water. It is therefore commonly found in the basin-environment clay facies of desert red-bed sequences, such as the Mercia Mudstones of England; it underlies some huge areas in North America (Ege, 1984) and is common in sedimentary basins worldwide. It may form relatively pure beds over 100 m thick, but is commonly interbedded with clays to form sequences sometimes known as Saliferous Beds.

Salt solution

A cubic metre of water can dissolve 360 kg of salt, to create a brine with a specific gravity of 1.23. Any assessment of salt solution by groundwater must also consider the solution rate and the available initial fracture size (James and Kirkpatrick, 1980), as well as the available through-flow of fresh water and brine. In real terms a cavity of significant size, nominally a conduit a metre in diameter, can be formed naturally in salt within a few years — in contrast to the time scales of around 10000 years needed to create a similar cavity in limestone. But large natural cavities are rarely stable in salt, due to the low mechanical strength of both salt (UCS around 10 MPa) and the normally associated clay rocks. Salt solution is usually accompanied by progressive collapse, creating micro-cavernous collapse breccias of insoluble residue, and causing gentle but continuous surface subsidence.

The solution process is so rapid that salt cannot normally exist at the ground surface; the exception is in desert climates, such as at some locations within the arid interior zones of Asia. Elsewhere, subsurface salt may only reveal its existence by the presence of natural brine springs. Where the geological structure dictates that salt should lie at outcrop, or at rockhead, there is usually just a zone of collapse breccia, immediately beneath any cover of soil or drift. These breccias may be over 100 m thick, and their evolution may have promoted tens of metres of ground subsidence (Figure 7.1).

Salt solution may continue beneath the breccia, but the salt itself is largely impermeable. Where surrounded by impermeable clays, ponding of the high-density brine on the salt–breccia interface is a natural limit to ongoing solution and subsidence. This restraint is, however, removed where outflow of the brine, and replacement by further fresh water, is possible due either to the regional geology and topography, or, very significantly, to artificial brine pumping. In addition to the regional subsidence over the entire salt outcrop, there is more severe local subsidence where groundwater flow is concentrated in brine streams due to variations in the breccia permeability and also where surface soils may slump or collapse into small salt cavities. Deeper circulation of fresh water and brine may also cause solution of buried salt beds, with subsequent upward stoping to form breccia pipes and surface subsidence remote from the salt outcrops (Figure 7.1).

Long-term salt subsidence

The Cheshire Plain of northwest England is underlain by thick beds of salt and mudstone, both overlain by a continuous cover of glacial drift. Where the salt has been in contact with the permeable drift, it has been dissolved by groundwater to leave collapse breccias which average 50–70 m in thickness (Evans, 1970; Earp and Taylor, 1986), whose contact with the underlying salt is locally known as the 'wet rockhead' (Figure 7.1). The age and rates of this solution, breccia formation and ground subsidence are unknown, but have extended through much of the Pleistocene. Some of the linear and circular subsidence depressions formed over the salt now contain some of the lakes, or

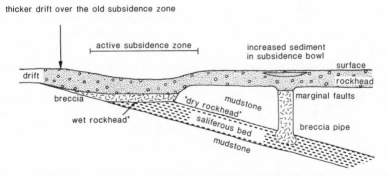

Figure 7.1 Profile through both the active subsidence zone over a salt outcrop and a breccia pipe developed by a long period of solution and collapse over a deeply buried salt bed. The saliferous bed is a sequence of salt beds and mudstones.

meres, which are a conspicuous feature of the Cheshire lowland. In many parts of Cheshire, the wet rockhead is 60–120 m below the surface, and lies far below the existing brine–freshwater interface. Salt solution at these depths appears to have occurred when this interface was depressed, due to high water tables within wet-base glaciers of the late Pleistocene (Howell and Jenkins, 1977); since the glacial retreat, lower water tables, thinner freshwater lenses and higher brine levels have retarded further solution, and surface subsidence rates have therefore declined (until increased by artificial brining). Similar collapse breccias are recognized where a thick salt bed comes to outcrop in parts of Kansas (Walters, 1977); there, the breccia is now covered by up to 90 m of Pleistocene alluvium, which fills the subsidence bowl created by the salt solution.

Even longer time scales lie behind some massive salt subsidence features in Saskatchewan, Canada. One of these is the Rosetown Low—a structural basin over 20 km across and 115 m deep; the salt was removed over 70 million years ago, and the basin is now filled with sediment (DeMille et al., 1964). Under this and other nearby structural depressions all the salt has been removed, and there is therefore no modern subsidence. But flowing brine springs do indicate that continued solution provides a subsidence hazard over some parts of Saskatchewan's enormous salt deposits.

A younger Saskatchewan salt feature is Crater Lake, whose sediments indicate that a zone 210 m in diameter subsided over 15 m during the last Pleistocene deglaciation. Geophysical surveys suggest that this subsidence overlies a classic example of a breccia pipe which extends to the top of the salt bed at a depth of 900 m (Gendzwill and Hajnal, 1971). The collapse breccia within this pipe exhibits a bulking factor of only 5.3%, suggesting it involved large block movement on clearly-defined ring faults. Other breccia pipes over salt beds at similar depths are known in the Hannover region of Germany. Clearly, subsidence on these features is well beyond man's engineering control, but the long time scales involved make the hazard potential insignificantly low.

As a crystalline rock, salt is distinguished not only by its high solubility but also by its high ductility, and it is subject to lateral flow under differential loading within a deep rock sequence. Salt domes are the most obvious consequence of this flow, and many of those beneath the Gulf Coast region of Texas and Louisiana are surrounded by annular zones of subsidence. Measurable modern movements on these are, however, probably more related to solution than continued lateral flow.

Modern natural salt subsidence

The commonest modern subsidences in the Cheshire saltfield are linear troughs, generally known as linear subsidences (Calvert, 1915). These are surface features up to 8 km long, but usually only around 200 m wide and 10 m

Figure 7.2 Moston Flash, a linear subsidence occupied by a lake in the Cheshire saltfield. The lake has formed from almost nothing within a period of 70 years, and the road across it has required regular raising, especially when the subsidence rate was high due to local brine pumping.

deep. They cut across the lowland region, breaching the low watersheds. They are active enough to warp Holocene alluvia yet also contain many peat deposits; they have developed continuously over the last few thousand years. The linear hollows have formed where solution of the salt is locally accelerated along brine streams. Also known as brine flows or brine runs, these are zones of concentrated groundwater flow along the salt–breccia interface (the wet rockhead), which is mostly at depths between 50 and 120 m below ground level. Many of the linear subsidences follow the strike of the salt beds, tracing the purer salt units within the variable salt and clay sequence of the Saliferous Beds (Evans *et al.*, 1968; Earp and Taylor, 1986), and others may follow faults. A prime site for active subsidence is along the edge of a buried salt outcrop where there is a greater supply of solutionally aggressive fresh water draining off adjacent clay, and this process leads to an asymmetric profile across the linear subsidence (as in Figure 7.4). These brine streams drain to the brine springs, which were common in Cheshire before the advent of brine pumping, but the most active subsidence zones are where the fresh water first meets the salt at the head of the brine stream, often far from the springs.

The natural subsidence rate within the linear features is low, and is often masked by the effects of brine pumping. Ground movement of 15 mm/year is an unusually high rate recorded along parts of a linear subsidence tracing the edge of a salt outcrop at Droitwich, in the English Midlands, ten years after brine pumping ceased in the area. Brine springs now feed into the local river.

The nature and permeability of the Cheshire plain drift vary considerably, but the linear subsidences are largely in areas of clay cover (Evans *et al.*, 1968).

Figure 7.3 The Meade sinkhole in Kansas, photographed 18 years after its overnight collapse above a buried bed of salt (photo: W. D. Johnson, United States Geological Survey).

Where the drift includes more sand, non-linear depressions up to a kilometre across progressively develop (Earp and Taylor, 1986), and smaller subsidence craters increase in numbers (Howell and Jenkins, 1977). The craters may develop within a few days, and Howell and Jenkins (1985) suggest these are due to the fluidization of low-density sands which are tapped off below into cavity networks in the breccia and salt; this mechanism is distinct from the downwashing of sediment which creates the normal subsidence sinkholes in limestone areas (see Chapter 3).

Perhaps because they form more rapidly and spectacularly, non-linear subsidences are also reported from many other saltfields; Ege (1984) reviews various examples in America. In 1970, a collapse sinkhole reactivated one of several Holocene depressions over breccia pipes 60 m in diameter above the salt beds in the Keuper Hills near Hannover, Germany. In 1879, the Meade sinkhole collapsed overnight in previously level ground in Kansas. It was 50 m across and half as deep, with a water table just 4 m below the surface, and another similar hole at nearby Rosel engulfed a railway station and buildings. These both developed due to salt solution at a depth of around 150 m, with subsequent piping failure and brecciation of overlying rocks and drift; a significant feature of the area is the juxtaposition of the salt with permeable sand beds which allows deep groundwater circulation (Frye and Schoff, 1942).

The complexity of the processes within the salt, breccia and drift, combined with the limited data base on natural subsidences, makes these effectively impossible to predict or control. Fortunately they are remote enough, in time and space, to offer a minimal hazard to engineering activities.

Man-induced subsidence over salt

Rates of ground subsidence over salt may be greatly increased above their natural levels by artificial changes in the patterns of brine and fresh groundwater flow. These are primarily achieved by brining—the commercial pumping of brine groundwater from the salt beds. This may involve removing the natural protective layer of brine which lies on wet rockhead; this brine layer is normally only a few metres thick, but when it is removed fresh water gains access to the salt, and solution continues. Alternatively, brining from deeper levels within the impermeable salt beds is possible with artificial injection of fresh water to dissolve the salt.

Ground subsidence induced by brining may be very localized and vertically above the sites of brine extraction; the geometry of these events relates to the subsidence profiles created by conventional mining and subsequent roof failure. On the other hand, brining can promote subsidence at considerable distances from the extraction site where the pumping is from natural long-distance brine streams, and this is effectively just an acceleration of the natural linear subsidences.

Linear and areal subsidences

The acceleration of natural salt solution, creating both linear and areal subsidences, is a notable feature of the Cheshire saltfield. The enhanced ground movements are due to 'wild brining', which is specifically the pumping of brine from boreholes sunk into natural, flowing, underground brine streams, the courses of which are often recognized by the natural linear subsidences already features of the landscape. For over 200 years the wild brining in Cheshire has caused these subsidences (Calvert, 1915); sometimes movement is as fast as a metre per year, and sudden flooding of subsidences, when they captured a river, created lakes locally known as flashes.

Moston Flash is a prime example of an active linear subsidence. It is 3000 m long, and its cross profile (Figure 7.4), about 200 m wide and 3–10 m deep, is remarkably uniform along its length. It has a 60 year history of conspicuous movement, and in the 1970s was subsiding 77 mm/year at a land-based post, and more in the centre of the flash. The asymmetric subsidence with slip scars on one side, the flash growth, and damage to roads and buildings are all typical, and relate to a brine stream on a pure bed of salt at the wet rockhead (Figure 7.4). By correlation over time of the volumes of pumped brine and ground subsidence, Oates (1981) has shown that the Moston Flash subsidence is due to wild brining at a well 2 km further north (Figure 7.5), subsidence

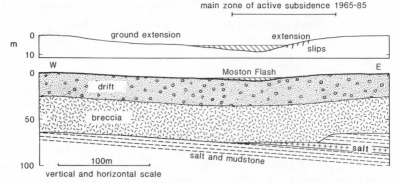

Figure 7.4 Cross-section through Moston Flash in a linear subsidence in the Cheshire saltfield, England. A brine stream is flowing through the breccia where it overlies a bed of massive salt at the wet rockhead, and solution and subsidence along this have been greatly accelerated by wild brine pumping. The brining ceased locally in 1978, but the ground movement on the east side continues at a lower rate. The upper surface profile has the vertical scale exaggerated by a factor of two.

lagging about a year behind the brining. Furthermore, rainfall in the flash is inadequate to supply the brine, indicating that fresh groundwater is drawn laterally through the drift and breccia onto the salt rockhead. The main zones of active subsidence migrate along the linear feature rather unpredictably; ridges in the rockhead are dissolved away to create localized subsidence, while rockhead hollows stabilize beneath ponds of protective saturated brine.

Distinct from the linear features are the areal subsidences, also described as irregular by Oates (1981). Elton Flash provides a good example (Figure 7.5) where around 60 ha of land have subsided a total of 9 m in the period 1892–1982. Maximum subsidence rate has been 700 mm/year and the centre of subsidence has shifted through time. This type of feature is probably due to fresh water draining down through a more permeable sandy drift to meet the salt at rockhead over a wide area, and then be drawn away as brine to a nearby wellfield (Figure 7.5). Where areal subsidences are more distant from the wellfields, Oates (1981) recognizes that more linear features develop within them as brine streams become established. Subsidence at both Moston and Elton Flashes is now greatly reduced as the relevant wild brining operations have recently ceased operation.

A third subsidence type is recognized by Howell and Jenkins (1977) as crater subsidence. This is a small sinkhole feature rapidly created when sandy drift flows into underlying cavities in the salt. Craters occur more frequently in areas near wild brining operations, but there is no field evidence that they coalesce into linear subsidences, though they may occur within them.

Localized subsidences over brining operations

Most modern brining operations use deep boreholes and injected fresh water to abstract salt from dry sites beneath an impermeable cover of clay or

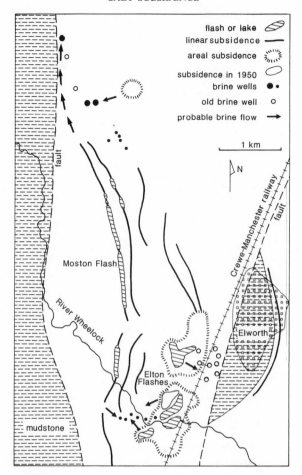

Figure 7.5 Linear and areal subsidences between Elworth and Middlewich in the Cheshire saltfield. Bedrock beneath the drift cover is the Triassic Saliferous Beds except where indicated as mudstone. The rapid development of Moston Flash was probably related to abstraction from the most northerly brine well, but pumping has now ceased at that site (partly after Oates, 1981).

mudstone. These sites are often described, rather incorrectly, as lying beneath a 'dry rockhead' (Figure 7.1). As there is normally no natural groundwater flow, the size of solution cavities created may be controlled to some extent. The controls do appear to be inadequate on some operations, as severe ground subsidences have occurred, and these demonstrate a number of different mechanisms of failure.

Brining at Hengelo, in southern Holland, exploits salt lying beneath 120 m of drift and plastic clay and 205 m of more competent mudstone (Wassmann, 1979). Each production well creates an inverted conical cavity about 45 m high and wide, but these flare out and are interconnected at the top of the salt bed; two subsidence bowls each over 2 m deep have formed over the largest

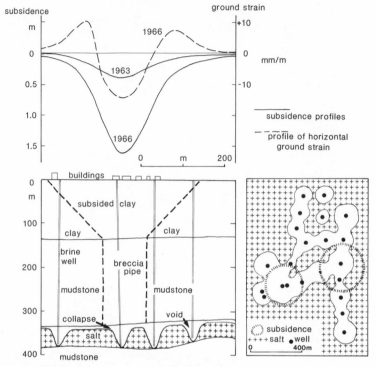

Figure 7.6 Subsidence profile and inferred cross-section of one of the major subsidence bowls developed over the buried salt in response to brine pumping at Hengelo, Holland. The inset map shows the location of the brine wells, the probable form of the interconnected cavities developed in the salt, and the two surface subsidence bowls above them (after Wassmann, 1979).

coalesced cavities (Figure 7.6). Profiles of ground strain and subsidence are similar to those over coal mines (Chapter 5), but the time scale is longer. At the second subsidence, there was no movement for five years after brining ceased on the central well; rapid movement then dropped the ground 1600 mm in three years, and 15 years later the annual subsidence was still 50 mm. Post-subsidence boreholes revealed piping failures which had migrated upwards into the mudstone and had then been followed by plastic deformation of the upper clays. The mudstone piping was a slow process as the dry rock was strong, and it only failed as brine worked its way upwards bed by bed into sag fractures and progressively weakened it. Observation of the first subsidence allowed successful prediction and engineering precautions when the second one developed (Wassmann, 1979).

Similar subsidences occur over brining operations in the Hutchinson salt which underlies a large area in Kansas, USA (Walters, 1977). Brining through unsealed boreholes creates large cavities at the top of the salt bed, and these coalesce to form galleries which may be hundreds of metres across with minimal remnant pillars. The gallery roofs sag and locally collapse, and surface subsidence ensues. At Cargill, gentle but persistent subsidence necessitated regular regrading of a railway line over two years, before a sinkhole collapse developed. Slow subsidence may have occurred at other sites in areas of farmland, but could have passed unnoticed.

Subsidence in the Kansas saltfield has also been caused by dissolution of the deep, dry salt beds following water ingress due to casing failures in abandoned oil wells (Walters, 1977). The salt is mostly at depths of 100–300 m and is breached by about 80 000 oil and gas wells drilled to reservoir rocks at depths of around 1000 m. Solution occurs either by undersaturated reservoir brines disposed of in old wells, or where the abandoned leaky wells permit groundwater circulation across the salt; subsidence due to the former ceases when the operation closes down, but in the latter case may continue for long periods. Significant movement has occurred at just eight wells, so it only rates as a rare event. But at Gorman, a subsidence bowl 300 m across is now 8 m deep, and continues to depress at 150 mm/year, providing an ongoing problem for the maintenance of a major highway across it.

Sinkhole collapses over salt

Rapid or instantaneous ground failures, creating large and deep sinkholes, occur where either the roof of an underground cavity collapses or un-consolidated surface sediments are washed down into a void. These are ubiquitous features of limestone regions (Chapter 3), but they may also occur in certain situations over salt.

Just north of Northwich, in the Cheshire saltfield, salt lies at depths as shallow as 30–40 m with a cover of only breccia and sandy drift (Calvert, 1915). Largely in the 19th century, numerous mines were sunk to the salt; they were only small but had extraction ratios sometimes as high as 90%. When abandoned, water entered through the shaft linings within the sub-water-table drift and started to dissolve the walls and pillars of the mine. Eventually either the whole mine collapsed, or the brine became saturated and solution ceased. Then in the period 1850–1930, 'bastard brining' was practised whereby brine was pumped from the old mines; as a result more fresh water flowed in, solution recommenced and failure became inevitable. The Northwich area became pocked with sinkholes mostly around 10 m deep and 100 m across, due to both mine collapses and the bastard brining. The collapses were so abundant that many coalesced and flooded to form expanding flashes, as illustrated in the map sequence by Douglas (1985).

The Cargill sinkhole is one of six collapses that opened up in subsiding areas over the Hutchinson Salt of Kansas (Walters, 1977; Dyni, 1986). Brining from beneath dry rockhead had been active at the site for 86 years, and flow patterns indicate that a number of cavities had coalesced into a single gallery with minimal support over a span of 400 m (Figure 7.8). Gentle surface subsidence over two years preceded a collapse, directly beneath some railway tracks;

Figure 7.7 The Cargill sinkhole which developed over a major brine pumping operation at Hutchinson, Kansas, in 1974. This photograph was taken only a few hours after the main subsidence; the railway tracks are left hanging in space, and a temporary fence keeps onlookers at a safe distance. In the next few days, the sides slumped in and the sinkhole more than doubled in area (photo: Hutchinson News).

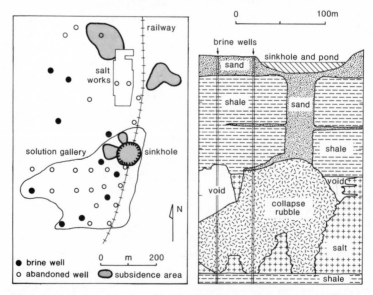

Figure 7.8 Area map and cross-section of the Cargill sinkhole which collapsed in 1974 over brine-pumped solution cavities in thick salt beneath Hutchinson, Kansas (after Walters, 1977).

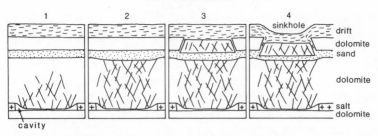

Figure 7.9 The possible sequence of processes leading up to the development in 1970 of major sinkholes over brine-pumped salt beds at Grosse Ile, Detroit. In stage 1, the middle dolomite sags and develops dilation fractures over the cavity in the salt. In stage 2, the overlying sandstone crushes to sand within the compression arch, and the sand runs down into the dilated dolomite fractures. In stage 3, the upper dolomite fails as a block, dropping into the void left by the migrated sand. In stage 4, the drift sediment runs into the fissures around the sunken dolomite block, and the surface subsides in a large sinkhole (after Stump *et al.*, 1972).

within three days, the sinkhole was 100 m across and 18 m deep. Extensive post-collapse drilling allowed interpretation by Walters (1977) of the sinkhole structure as in Figure 7.8. A chimney 30 m in diameter had developed in the shale over the salt by stoping and collapse, with bedding plane dilation fractures also formed around it; this had eventually reached the un-consolidated sand which had then run-in freely.

The size of the Cargill sinkhole was a function of the run-in of the loose sand overburden, but a different mechanism has been ascribed to three massive sinkholes at Grosse Ile, Detroit (Ege, 1984; Nieto and Russell, 1984). Brining from a depth of 340 m has created massive open galleries beneath the island, and around 1970 the sinkholes, one 130 m in diameter and the other two half the size, formed slowly and consecutively over a period of months. Post-event boreholes did not reveal breccia pipes below the sinkholes, and the postulated sequence of events is as in Figure 7.9. The critical factor was the sandstone, with a tested strength of only 8.3 MPa, which crushed to sand when caught in the compression arch developing over the failing cavity roof. The sand then flowed into dilation fractures in the dolomite below, and allowed a block failure of the overlying dolomite, followed by collapse of the drift and ground surface—instead of just further gentle sag subsidence. The geology of the weak sandstone at Grosse Ile may be unusual, but there could be implications relevant to sinkhole hazards elsewhere.

Major sinkhole collapses have also occurred where abandoned, unsealed oil wells have created groundwater access to previously dry salt beds, as described above; cases are recorded in Kansas (Walters, 1977) and Texas (Johnson, 1987), and ground failure may lag many years after plugging of the well.

Mining precautions

Brining operations which create large underground cavities, whose extent and stability is uncontrolled and largely unknown, almost inevitably lead to ground subsidence. The only long-term remedy is to close down the inappropriate operations. Bastard brining from old mines, which once caused such major destruction in the Cheshire saltfield, has long been illegal. Wild brining causes less severe subsidence, and it cannot reasonably be banned in Cheshire because of its role in the industrial infrastructure of the region (Collins, 1971). It is, however, on the decline, and no new permissions are now granted. Many areas of subsidence have stabilized soon after nearby wild brining has terminated.

The old Cheshire situation was mimicked by the numerous small, shallow, wild brining operations at Borabu in Thailand, which promoted a series of sinkhole subsidences (Rau *et al.*, 1982); because of these, and also because of waste brine contamination of agricultural water, the brining was stopped by Royal Decree in 1980.

Ground subsidence does not occur over properly controlled brining operations beneath dry rockhead (Figure 7.10). At sites in Cheshire, water is injected through grids of single boreholes to create cavities over 50 m high and wide within the salt; these are monitored by downhole sonar, and roof dissolution is prevented by oil or air cushions; total extraction is only 20% and the cavities are not allowed to interconnect. An alternative system used in America employs hydrofracturing with high-pressure water, to create fractures linking pairs of wells 100–300 m apart; dissolution then creates long, narrow galleries with stable arched roofs (Ege, 1984). In both systems, the cavities are finally left full of saturated brine, or may be backfilled with inert chemical waste. Additionally, totally dry pillar-and-stall mining of salt (Chapter 6), beneath dry rockhead, is stable; 75% extraction rates of salt up to 8 m thick are achieved in Britain and America with no ground subsidence.

Continued wild brining in Cheshire justifies a compensation scheme for subsidence damage. Instigated in Northwich in 1892, the Cheshire Brine Subsidence Compensation Board has since 1952 covered the whole county saltfield. Funded by a levy on all pumped wild brine (and also a smaller levy on controlled brine operations), it pays compensation for damage to buildings

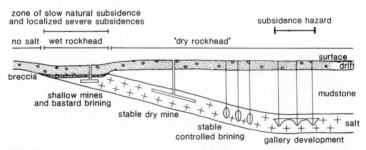

Figure 7.10 Subsidence hazards from contrasting methods of salt extraction by either mining or brining. Bastard brining from abandoned shallow mines is now illegal in Britain. Gallery development in deep brining may become unstable if cavities coalesce through lack of adequate control.

and structures (but not to the local council, railways or canals, which are deemed to profit from the salt industry). The Board offers advice on, and may make discretionary payments for, foundation and engineering precautions, and also carries out remedial works. Acceptance of its advice is a prerequisite for planning permission in designated subsidence zones (Collins, 1971).

Engineering precautions

Catastrophic sinkhole collapses over salt are generally so rare, in both time and space, that engineering precautions are unrealistic. The exception is in areas of shallow wet rockhead with irresponsible brining operations. Such was the case in the Northwich area of Cheshire, around the year 1900, where the subsidence destruction was perhaps the most extensive ever known. Some roads sank 10 m, and many buildings had to be demolished; people gathered in crowds to watch the big sinkholes develop in an area of over 100 ha which was totally destroyed (and is now covered by chemical waste lagoons). The chaos in the town was spectacular; houses leaned 0.5 m out of true, and some had 10 cm cracks in them; lakes formed 10 m deep, rivers sank into holes in their beds, and canals were emptied overnight; there was even loss of life. The descriptions by Calvert (1915) are classic (also Ward, 1900), and some are reviewed by Bell (1975) and Waltham (1978), while Wallwork (1960) describes subsequent dereliction. The only possible remedy was to stop the mining and brining.

Figure 7.11 One of the many cases of catastrophic sinkhole subsidence in the town of Northwich in the Cheshire saltfield, around 1900 when shallow salt mining and bastard brining were at their peak (from an old postcard).

Slower subsidence is more widespread, and can be tolerated by precautionary engineering design. Lower parts of the town of Droitwich, England, have
subsided 8 m, and much of the High Street area is now on 5 m of fill; along the
natural linear subsidence through the town, an 80 year building life must
anticipate a metre of subsidence and also rotation of 1 in 40. In Cheshire the
regional structural planning is designed to avoid the areas of wet rockhead
where possible (Collins, 1971). The patterns of subsidence tend to repeat over
time, and the known areas of severe movement, and also river-level areas, are
avoided entirely by new building. Prediction of subsidence has proved possible
on a local basis within the deep, semi-controlled brine field in Holland
(Wassmann, 1979). Wire-line logging of abandoned but accessible wells has
shown over 2 m of displacement over 30 months at depths of 250 m, but
progressively less movement towards the surface; there was no ground
subsidence at the time, but it could be anticipated in the future when bed
failure migrated upwards.

Subsidence precautions at Northwich, Cheshire, have seen the use of timber-
frame buildings since 1890. Some modern buildings have steel frames, but
these add 40% to the cost of houses. In the current era of more modest
subsidence, concrete rafts at half the extra cost are found to be appropriate.
Buildings with a frame or raft can then be jacked up, and one farm near
Moston Flash has been raised and levelled seven times since it was rebuilt in
1967. A single large house can be levelled, using 50-ton jacks, for less than
£1000. In the past, whole streets have been raised in Northwich to keep both

Figure 7.12 Timber frame buildings are dominant in the older parts of Northwich, in the Cheshire
saltfield. The frames were to allow the buildings to be jacked up to compensate for the extensive
salt subsidence.

Figure 7.13 Lifting shops along one of the main streets of Northwich around 1900. This occurred frequently at the time and was followed, or preceded, by the raising of the street to keep the town above the local river level in the face of rapid subsidence caused by uncontrolled brining operations in the salt beds less than 30 m below the ground (from an old postcard).

Figure 7.14 Active subsidence affecting the Crewe–Manchester railway where it crosses the salt outcrop at the Elton Flashes in the Cheshire saltfield. The track alignment shows the lateral ground movement into the subsidence bowl, and the adjustable gantry supports are to allow for reballasting after vertical movement.

roads and buildings above flood levels. Large structures are undesirable, and a
school in Droitwich is successfully founded on 14 independant jackable rafts.

Roads, railways, rivers and canals may require constant maintenance with
regrading and rebanking. Bridges should have built-in jacking points,
especially where they are over water and have to maintain clearance. The
Manchester to Crewe railway incorporates various measures where it crosses
the Elton Flash subsidence (Figure 7.5); it has a cellular bridge in its
embankment which is regularly raised, and frequent reballasting of the track
has necessitated the use of special adjustable gantries for the overhead power
lines. Reasonable design precautions are economically sound and, except in
the rare cases of sinkhole collapse, are adequate to maintain structural
integrity in any salt region where subsidence is a likely hazard.

Solution of gypsum and other rocks

Salt is not the only highly soluble rock, but other comparable materials, such
as sylvite, are too rare to pose a sensible subsidence hazard. Solution of
limestone is so slow that it does not produce significant ground subsidence,
except by the important secondary mechanisms of cavern failure (Chapter 2)
and sinkhole development (Chapter 3). Gypsum (hydrated calcium sulphate)
has an intermediate solubility 100 times that of limestone, but less than one
hundredth that of salt. It can therefore be dissolved in groundwater rapidly
enough to promote ground subsidence, yet it is stable enough and strong
enough also to form caves which may then pose a collapse threat.

Gypsum underlies much of the Ripon area in northern England; at depths of
over 100 m it is replaced by anhydrite, and at outcrop the horizons are mostly
brecciated mudstone, with restricted gypsum exposures revealing solutional
fissuring (Cooper, 1986). The interbedded rock includes permeable Magnesian
Limestone, and by analogy with the salt situation, there is wet rockhead over
an area much wider than outcrop. Deep natural solution of the gypsum is
indicated by collapse breccia pipes, 10–20 m in diameter reaching 100 m
through the overlying rocks, and gypsum is absent beneath one area of 8 km^2
where rockhead is depressed 25 m beneath a clay-filled hollow. Numerous
small sinkhole depressions occur over the gypsum, exhibiting surface
failure in both drift and rock. Their collapse mechanisms are considered in
Chapter 3. There are also larger depressions only a few metres deep and up to
300 m across (Cooper, 1986), where subsidence is gentle and locally active.
These may be considered as either the coalescing of many small sinkholes over
cavern collapses, or as a more continuous sequence of solution and collapse
comparable to the rockhead breccia formation on salt.

A similar situation of slow, shallow, areal subsidences, as well as smaller
collapses, pertains in a gypsum area near Zaragoza in Spain (Benito and
Gutierrez, 1987). At this site, many buildings and rocks have failed and one
new town development has been abandoned due to subsidence. At Ripon, the

Figure 7.15 A terrace of houses distorted by slow subsidence due to solution of gypsum below a shallow rockhead; in Ripon, Yorkshire.

subsidence activity approaches the total unpredictability of limestone sites; either existing depressions deepen further, or intervening ridges eventually fail (Cooper, 1986). On a factory site in one broad depression, foundation piles were driven to resistance at depths varying between 6 and 30 m, where rockhead is a complex of collapsed materials. Raft foundations are justified for many buildings, and site investigation needs to be very thorough.

8 Regional subsidence due to fluid withdrawal

Subsidence on clay soils as a direct consequence of the over-abstraction of groundwater from interbedded sand aquifers is a widespread phenomenon. Gentle subsidence bowls develop almost imperceptibly slowly but can extend over large areas. Their main effects are coastal inundation and deformation of surface drainage gradients, together with casing damage to the wells which initiated the subsidence and some cases of structural damage through ground strain. The Koto area of eastern Tokyo has subsided over 4 m since 1920 due to a pumped water-table decline of 60 m; two million people now live below high tide level, and massive flood defences and pumped drainage schemes have been installed, while an extensive landfill project is the only way to eliminate the flood threat posed by a major typhoon.

This style of induced subsidence has affected 40 areas in Japan, mostly round its coast, together with 18 areas in the southern states of the USA, and many more in geologically young terrains of other countries. On a much smaller scale, the same mechanism can promote destructive settlement around construction sites dewatered by well-pointing; this is most severe in normally consolidated clays, where it may demand special provision of drainage barriers (Powers, 1985). Although shallow groundwater abstraction is the main cause of the regional subsidence, an equal scale of ground movement can be achieved through the abstraction of oil, gas or geothermal fluids, though in these cases the subsequent subsidence is not ubiquitous.

The worldwide scale of subsidence damage through groundwater withdrawal reached a peak between 1950 and 1970, at a time of unprecedented urban growth and industrialization. An extensive literature on the engineering aspects has since accumulated, including three UNESCO-sponsored conference volumes (published by the International Association of Hydrological Sciences as Publications nos. 88–9, 121 and 151). It is significant that the first conference (at Tokyo in 1969) was dominated by case histories of the effects of water-table decline, while the third (at Venice in 1984) was more concerned with groundwater modelling and subsidence monitoring—many of its papers turned to other styles of subsidence. Within that period, the mechanisms of subsidence due to fluid withdrawal were largely understood, and the review edited by Poland (1984) is therefore a valuably complete statement on the subject.

Subsidence mechanism

Compression of aquifers in direct response to groundwater head decline has been recognized for over 50 years. It is a simple consequence of the increase in effective stress when the porewater pressure decreases. However, sand is almost incompressible under the stress of shallow aquifers, except for a small amount due to grain rearrangement. Only at the higher pressures pertaining in deep oil reservoirs does grain fracture promote further compression under increased effective stress. Lithified rock aquifers similarly exhibit only minimal compression on draining, but this may still be significant. The Zeuzier Dam, a double arch structure in Switzerland, performed satisfactorily for 20 years, and then, in 1979, developed cracks as it subsided rapidly for 80 mm followed by further slow movement. Even though the dam is founded on massive limestone, the subsidence was instigated by considerable under-drainage into the Rawil road tunnel then being constructed 1400 m away and 400 m below the dam crest (Egger, 1983).

The compaction of a sand aquifer in response to head decline is immediate and elastic, and is usually small. Far more important as a cause of major ground subsidence is the subsequent compaction of interbedded clay aquitards, which though not pumped themselves, suffer a similar increase in effective stress as porewater pressures equalize with the lower values induced in the adjacent aquifers. The wide distribution of sand–clay sequences beneath alluvial plains in many parts of the world accounts for the great extent of this style of subsidence, especially as the shallow sand aquifers are all too simple to exploit and over-pump.

This greater compaction of the clay aquitards is largely inelastic and so non-recoverable. The amount of resultant subsidence is a function of the coefficient of volume change (directly related to the hydrologists' non-recoverable specific storage), which is determined by the existing geology, and also the induced stress change due to water-table decline, which is both imposed and controllable by man (Helm, 1984). The specific storage is the amount of water which has to be squeezed out of the aquitard to achieve consolidation at the new increased level of effective stress, and this varies considerably with the clay mineralogy. It increases by a factor of three, from the little-compressible kaolinites, through the illites, to the highly compressible montmorillonites (the dominant and influential member of the smectite clay mineral group). A porosity decrease of just a few percent within aquitards is induced by drawdown on a scale which is not unusual, and easily accounts for the observed ground subsidence in overpumped sediment basins.

The head decline induces two stress changes within the aquitards. Gravitational stress is increased through loss of buoyancy in the drained clay, and seepage stress is induced by groundwater flow through the aquitards (Poland, 1984). The latter may be negative or positive as the clay drains upwards or downwards into the overpumped sand layers. The consequent compaction of the clay is slow due to its low permeability. Not only does this lag the start of the subsidence behind the aquifer head decline, as is widely recorded, but it also permits subsidence to continue after the head has ceased to fall; subsidence at Savannah, on the Atlantic coastal plain of the USA, continued for over 12 years after the aquifer head decline had ceased (Davis, 1987). Furthermore, the subsidence is only initiated (or rapidly increases) when the effective stress on the clay exceeds any preconsolidation stress created by overburden loading before erosion (or by an earlier cycle of drawdown); at Franklin, Virginia, 20 m of pumped head decline took place before ground subsidence developed (Davis, 1987).

Perhaps the type area which most clearly demonstrates this type of subsidence is the Santa Clara Valley, now often known as Silicon Valley, at the south end of California's San Francisco Bay. Subsidence has locally reached 4 m, and 44 km^2 of the Bay margin are now below sea level. Levée construction, road raising, well repairs and loss of land values have already cost over

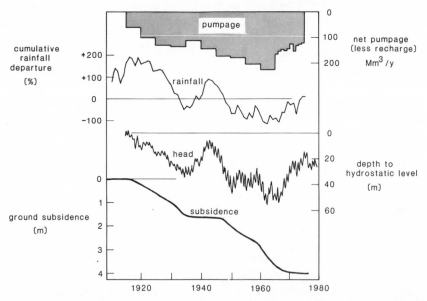

Figure 8.1 Ground subsidence recorded at a benchmark at San Jose, in the Santa Clara Valley of California for 1912–1976. The profile shows a clear correlation with the volumes of pumped groundwater (less recharge in the later years), rainfall (expressed as cumulative % departure from the 50 year mean) and the resultant decline of the water table recorded in a nearby well (after Poland, 1984).

$35 million (Poland, 1984). The valley is underlain by 1000 m of sediment, with clay aquitards rich in montmorillonite forming 25–80% of the succession. Pumping from the shallow interbedded sand aquifers, for agricultural use and then industry, caused a major head decline from 1912 to 1935 (Figure 8.1), accelerating as rainfall fell below normal in the later years. Subsidence of over 1 m matched the head decline, but stopped when the water table recovered in the wet years of 1936–42. A second phase of head decline caused no subsidence until the water table fell below its previous low, and then subsidence increased until 1965, to reach a total of nearly 4 m (Figure 8.1).

Protruding well casings in the Santa Clara subsidence bowl showed the importance of compaction of the shallow sediments. Extensometers in some old wells then proved that nearly all the compaction occurred in the top 300 m of sediments—which included all the pumped sand aquifers. Within this depth the total aquitard thickness is 145 m with a mean porosity of 37%, and the 4 m of subsidence is due to the porosity decreasing to 35.2%. With 3.84 m of subsidence from 49 m of head decline, the specific subsidence was 0.08 (metres per metre of head decline), but the compaction was incomplete due to the low aquitard permeability, and Helm (1977) estimated the ultimate specific subsidence would be 0.11, causing 5.3 m of subsidence for the same 49 m head decline. Dividing by the aquitard thickness, the virgin compressibility of the Santa Clara clays is therefore $7.4 \times 10^{-4}/m$—a figure close to those of many subsiding clay regions.

Subsidence prevention

The temporary halting of the Santa Clara subsidence in the late 1930s, when the run of wet years naturally raised groundwater levels, demonstrated the means to control and prevent subsidence due to drained clay compaction. Stabilization of the water table greatly reduces the subsidence rate, and head recovery to its initial level halts the subsidence far faster than it was initiated by the corresponding head decline (Helm, 1984). The Santa Clara Valley's second phase of subsidence, starting in 1948, was prematurely interrupted and brought under control by the water-table rise starting in 1967, and ground movement had virtually ceased by 1973. The head recovery was partly due to increased rainfall, and partly to imposed measures (Figure 8.1); pumping was decreased by groundwater taxation, surface water was imported via new aqueducts, and surplus imported water was used to recharge the unconfined aquifers by infiltration (Poland, 1984).

Various remedial measures are applicable to halt this style of subsidence. Much of Tokyo's subsidence was almost stopped by 1965, when a head decline of 60 m was replaced by recovery of 30 m through legislation to reduce groundwater pumping. Aquifer recharge is possible by feeding imported water to influent streams or leaky retention basins on unconfined aquifers, and a recharge tank has been designed and evaluated to return water to the shallow

E

confined aquifer in Bangkok (Nutalaya *et al.*, 1986). Water injection into the aquifers through existing wells has been successfully carried out in some subsiding oil and gas fields (see below), and Shanghai's subsidence was stopped when river water was injected into shallow overpumped aquifers. The early stages of the piecemeal drainage of Holland's Ijsselmeer, to form land polders below sea level, caused subsidence of adjacent land, so Flevoland, drained in 1968 (de Glopper, 1986), was made as an island with a wide marginal channel under which groundwater pressures could recover. But the Markerwaard polder is underlain by a confined aquifer whose head will decline 5 m on drainage, and its overlying aquitard will minimize the effect of a marginal channel (Koning, 1986); subsidence prevention has been shown to be possible and economic (Vos *et al.*, 1986) by letting the surrounding water levels recover through either a line of recharge wells or a maintained infiltration groove cut through the aquitard in the floor of the channel. The clear relationship between subsidence and groundwater pumping has meant that the abstractor can now be held legally responsible for the subsidence damage, in at least some of the states of the USA (Carpenter and Bradley, 1986).

A further consequence of head increase can be the recovery of the small amount of elastic compaction. This creates surface rebound when water tables are raised to, and maintained at, close to their original levels, but the rebound is normally only a few percent of the previous subsidence. Following 2.37 m of subsidence at Shanghai, head recovery permitted 34 mm of rebound, and

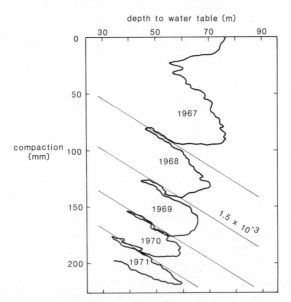

Figure 8.2 Correlation of vertical stress (recorded as the depth to the water table) with compaction (and hence subsidence) through five years at a site in the Santa Clara Valley, California. This shows that the recoverable compaction, or the coefficient of elastic deformation, is 1.5×10^{-3} metres per metre of head change (after Poland 1984).

rebound on a similar or larger scale has also been recorded at Tokyo, over the Wilmington oilfield in California, and at Venice (see p. 123). In the Santa Clara Valley, net rebound of around 10 mm occurred in 1974 before interruption by climatically induced water-table changes (Poland, 1984). Seasonal oscillations at Santa Clara allowed correlation of head changes and ground movements (Figure 8.2), where the recoverable components are recognizable when water levels are high; the coefficient of elastic deformation can be read as 1.5×10^{-3} for the complete system, or 6.15×10^{-6} per metre of sediment (Poland, 1984). This represents about 0.8% of the virgin compressibility.

Classic cases of subsidence due to water abstraction

So many of the world's major cities have suffered subsidence self-induced by groundwater pumping that there is an extensive literature of case histories (including a useful collection in Poland, 1984). Subsidence at Tokyo, Osaka and some other Japanese cities has been controlled by reducing pumping (Yamamoto, 1984). At Shanghai the subsidence has largely been stopped by aquifer recharge, and continuing time-lagged compaction of the aquitard clays is now partly matched by rapid elastic rebound of the aquifer sands (Su, 1986). London too has subsided, by up to 350 mm, due to 100 m of head decline in its chalk aquifer (Wilson and Grace, 1942), but the specific subsidence is low, as the compressing London Clay has a high level of preconsolidation stress.

The most instructive case studies are those of California's Santa Clara Valley, already referred to, and the San Joaquin Valley in the same state, together with the cities of Mexico, Venice, Houston and Bangkok, which are reviewed in the following paragraphs.

The San Joaquin Valley in California has the world's largest subsidence bowl, 9900 km² in extent and reaching 9 m deep. Superimposed within it are smaller bowls of subsidence due to hydrocompaction (Chapter 11) and oil extraction. Subsidence damage, mainly to wells and irrigation canals, and some necessary levée construction, has cost over $50 million and the whole case is well documented (Ireland et al., 1984; Poland, 1984). Thick sediments underlie the valley and the top 600 m, which record nearly all the compaction, contain 50% of compressible aquitards rich in montmorillonite (Meade, 1967). Over 100 000 wells exploit the intervening aquifers, and water tables have declined locally by over 150 m. Groundwater modelling has revealed massive vertical water transfers through long runs of perforated well casings, and also an incomplete subsidence profile (Prudic and Williamson, 1986). Over time, the volume of subsidence has correlated with about one third of the volume of abstracted water, and can be accounted for by a porosity decline of 1–4% in the aquitards (Poland et al., 1975). With geological variations present, the specific subsidence ranges from 0.01 to 0.08. Areas with extensive sands and thin clays initially subside faster, due to rapid water expulsion from the aquitards, and only when ultimate compaction is approached do the areas of

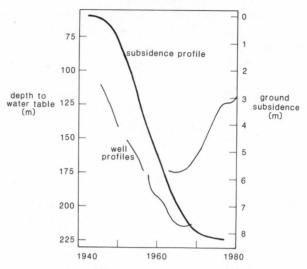

Figure 8.3 Subsidence at a benchmark near Mendota in the San Joaquin Valley, California, correlated with hydrographs from four nearby wells (smoothed to eliminate seasonal changes) for the period 1940 to 1980 (after Ireland *et al.* 1984).

thicker clays and lensing sands dominate the subsidence profile (Bull, 1973). Mean virgin compressibility is $7.5 \times 10^{-4}/m$, and the recoverable elastic compression is around 1.25% of this (Poland, 1984). Since 1970, water imports via the California Aqueduct have permitted reduced pumping, and head recovery has virtually stopped the subsidence (Figure 8.3) and introduced some limited rebound.

Water abstraction from beneath Mexico City has caused subsidence which has locally reached 9 m. The city lies in a basin largely floored by an old lake bed, and the ground conditions fall into three distinctive zones (Figueroa Vega, 1984). The city centre lies in the lake zone, where two shallow sand aquifers (Table 8.1) are interbedded with clays which are over 80% montmorillonite and have exceptionally high water contents (Zeevaert, 1972), and potential subsidence in this zone is > 10 m. The city suburbs spread on to the

Table 8.1 Generalized geological succession in the lake zone of Mexico City and location of soil compaction which causes ground subsidence (after Zeevaert, 1972)

Depth range	Lithology	Water content % dry weight	Contribution to total subsidence
0–5	fill		
5–32	silty clays	200–400	55
32–35	sand aquifer		
35–45	silty clays	100–200	30
45–62	sand aquifer		
> 62	clays and sands		15

transition zone of coarser alluvial fan sediments where potential subsidence is
< 10 m, and also on to the bedrock hills zone where subsidence is no threat.
Over 3000 wells into the shallow aquifers, originally artesian, together with
200 wells into aquifers 100–300 m deep, have abstracted water far in excess of
recharge, so that heads have declined and the subsidence has been induced.
The compaction of the shallow sediments has been dramatically demonstrated
by the protrusion of well casings, some of which have risen 5 m above the city
streets. Monitoring has shown that most of the subsidence is due to
compaction of the top two aquitards (Table 8.1). Imported water supplies and
some controls on pumping have promoted partial head recovery, and the
subsidence rate has decreased, from local maxima of 900 mm/year in 1952, to
around 30 mm/year (Figueroa Vega, 1984).

   The infamous subsidence of Venice has caused increasingly frequent
flooding of the unique sea-level city, driving out the permanent population
and threatening the survival of the splendid heritage site. Within the city of
Venice, ground subsidence has only reached a maximum of 150 mm during
this century (Cartognin et al., 1977), though it has been more at the industrial
zone of Marghera, next to the landward end of the lagoon causeway. Venice
stands on 1000 m of sediment, with six sand aquifers interbedded with clay
aquitards in the top 350 m (Gambolati et al., 1974). For centuries there has
been natural subsidence at a rate which is now 0.4 mm/year (Carbognin and
Gatto, 1986) due to both crustal sagging and sediment compaction. But since
1925 there has been additional induced subsidence due to groundwater
pumping, mostly from wells at Marghera. By 1969, the average head beneath
Venice had declined 9 m, causing an extra 120 mm of subsidence (Figure 8.4);
head decline and subsidence were greater at Marghera but of little environmental impact. The specific subsidence at Venice was 0.013, though this was

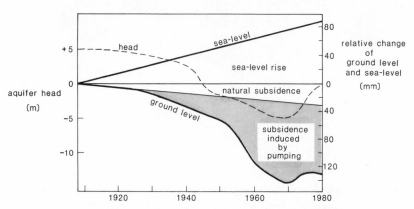

Figure 8.4 Ground subsidence, both natural and artificially induced, and sea-level rise at Venice
accumulated to cause 220 mm of additional flooding through the period 1908–69. Head recovery
since 1969 caused some rebound, the slow subsidence continues (after Gatto and Carbognin,
1981).

Figure 8.5 The subsidence of Venice is demonstrated by an acque alta in 1984 which forces pedestrians to use raised broadwalks to cross the flooded Piazza de San Marco (Photo: Associated Press).

not an ultimate value. At the same time, worldwide sea-level rise has been 1.27 mm/year. The combination of these factors meant that Venice sank 220 mm relative to sea level in 1908–69 (Figure 8.4), at a rate for greater than the earlier natural movements. Consequently the frequency of spring high tide flooding (acqua alta) increased from once per decade around 1900, to nearly five times per year at present, and the 1966 floods caused damage to the extent of $85 million. Severe pumping restrictions, with a new aqueduct for imported supplies, reversed the head decline in 1969, rapidly promoted 20 mm of rebound (an unusually high proportion of the initial subsidence, probably due to the low content of inelastic montmorillonite), and let subsidence return to its low, natural, unstoppable rate. But the periodic flooding of Venice continues. Remedial schemes to repressure the aquifers inside a cut-off wall, or for massive grout injection (Gallavresi and Rodio, 1986) are impractical for the city. The currently favoured protection scheme, to install lifting barriers to close the lagoon entrances during any acqua alta, is estimated to cost over \$2500 million and is unlikely to be operative before the end of the century.

Subsidence in the Houston and Galveston area of Texas has exceeded 0.5 m in a bowl 70 km across (Gabrysch, 1984), and has reached a maximum of 2.3 m. This is due to over-abstraction of groundwater lowering the average head by up to 90 m in a complex of sand aquifers and clay aquitards. Costs of the subsidence, mostly through coastal flooding, levée construction, drainage and falling property values, exceed \$200 million. Since 1975, pumping controls and imported supplies have produced some head recovery and the subsidence has slowed, and computer modelling of the groundwater has yielded subsidence predictions reaching 50 years ahead (Neighbors and Thompson, 1986). A notable feature of the Houston subsidence has been the activation of faults within the sediments, which have localized surface movement and increased the scale of structural damage. The faults predate the subsidence, so their distribution does not relate to the subsidence bowl, but they have been activated either by ground tension in the outer zones of the bowl, or by differential compaction especially where the fault has acted as a hydrological barrier (Kreitler, 1977). Comparable features have been recorded in subsidence bowls in Arizona (Holzer, 1984). Long earth fissures have opened by a few centimetres, under tensile ground strains around 0.0019 (Larson and Pewe, 1986), and fault scarps have grown to heights of 0.5 m where they coincide with the maximum slopes created in the subsidence bowls.

Bangkok is worthy of note as it is probably the city with the highest current rate of uncontrolled subsidence. Furthermore it houses four million people at ground elevations only 1–2 m above sea level (Akagi, 1979; Bergado et al., 1988). Pumping from eight aquifers has caused over 50 m of head decline, and subsidence has locally reached 1.2 m already (Nutalaya et al., 1986). The major pore pressure decline has been in the soft Bangkok Clay, at depths of 10–20 m and directly above an overpumped sand aquifer. But salt water intrusion has prompted exploitation of deeper aquifers, and 60% of the subsidence is now due to aquitard compaction at depths greater than 50 m. Future subsidence is likely to exceed 2 m (Nutalaya et al., 1986), putting much of the city below sea level, and already the low gradient storm drains have lost efficiency across the subsidence bowl. Remedial action, with groundwater taxation, pumping controls, imported supplies and aquifer recharge, is becoming increasingly urgent.

Controlling factors and prediction of subsidence

It is clear from the many case histories that regional ground subsidence, at any one site, is directly related to the pumped decline of groundwater head. But this relationship, the specific subsidence, varies between sites in response to geological factors. Of prime importance is the aggregate thickness of compressible clay aquitards interbedded with the overpumped aquifers; in the San Joaquin Valley of California the specific subsidence rises from .01 near the basin margins to .08 in the centre, as coarse alluvial fans give way to finer distal sediments (Bull and Poland, 1975). In the Houston area, Gabrysch (1970) found the specific subsidence increasing linearly from .005 where clays occupied 40% of the drained sediment sequence to .02 where there was 70% clay.

The third controlling parameter is the montmorillonite content of the clays, as this mineral exhibits uniquely high compressibility. The role of montmorillonite is widely recognized (Poland, 1984; Davis, 1987), and is broadly demonstrated by Table 8.2. Though the data in this table have been meaned or generalized to allow comparisons, the virgin compressibilities of the clays at Venice, Houston, Santa Clara and Mexico, all geologically recent, normally-consolidated sediments, clearly reveal the role of montmorillonite. Furthermore, the data from London and Savannah perhaps reflect the same influence, but also show that older preconsolidated clays have far lower compressibility, and the geological age and history of the clay must rank as the fourth major control on induced subsidence.

Further interpretation from Table 8.2 is unreal because of the limited range of generalized data, and mathematical relationships of subsidence to individual parameters are hindered by the many variations between sites. Besides head decline, clay thickness, montmorillonite content and preconsolidation history, a complete recognition of the subsidence process must also involve the proportions of other clay minerals and silt, diagenetic evolution, and the progress towards ultimate compaction through time.

Prediction of subsidence is therefore only feasible for an individual site where these various parameters are known or remain constant, and the available techniques are reviewed by Helm (1984) and Poland (1984). Purely empirical prediction methods are of limited value as a time–subsidence correlation is rarely linear. More successful is a calibration of subsidence against a single parameter, either head decline or withdrawn volume, which makes correlation and reasonable prediction possible for any single site. The alternative is a completely theoretical approach based on consolidation theory (Poland, 1984). Helm (1984) takes the theoretical approach further with two prediction schemes: a simple depth–porosity model is applicable in a new area with limited available data, while an aquitard drainage model is more revealing and more reliable with respect to time and recovery, but relies on data accumulated as the subsidence evolves.

Table 8.2 Comparison of geological, hydrological and subsidence parameters at six sites, with data taken from various published sources referred to in the text. Subsidence at Santa Clara is quoted as the predicted ultimate value which has not been realized; all other subsidence figures are those recorded, which are not ultimate.

Site	Clay thickness m	Head decline m	Subsidence m	Specific subsidence m/m	Compressibility $\times 10^{-4}$ m/m/m	Montmorillonite content %	Age
London	60	100	0.35	.0035	0.6	0	Eocene
Savannah	50	48	0.19	.004	0.8	60	Miocene
Venice	130	9	0.12	.013	1.0	10	Recent
Houston	150	90	2.3	.025	2.1	50	Recent
Santa Clara	145	49	5.3	.11	7.4	70	Recent
Mexico	50	55	9.0	.16	32.0	80	Recent

Withdrawal of oil, gas and geothermal fluids

Subsidence over producing fields of oil, gas or hydrothermal fluids is commonly restricted by the low compressibilities of the lithified rocks at the great depths of most fluid reservoirs. In a review of oil and gas fields, Martin and Serdengecti (1984) describe subsidence as rare and usually small, but there are cases where severe ground movements have occurred. Another contrast with the groundwater situations is that the deep abstraction of pressurized energy resource fluids involves higher stress of the rocks and most of the subsidence is due to compaction of the reservoir rocks, frequently by grain fracture in the sandstones, while aquitard compaction is less important.

Over 4 m of subsidence has been induced over some of the Venezuelan oilfields around Lake Maracaibo, where the yielding reservoirs are at depths of 300–1200 m (Nuñez and Escojido, 1977). Initially subsidence was very slow, until reservoir stresses passed preconsolidation thresholds, and then it showed a linear relationship with the abstracted oil volumes, except where distorted by lateral oil migration. The subsidence was due to compaction of both the reservoir sands and the interbedded clays as fluid support was withdrawn.

The greatest recorded oilfield subsidence is at Long Beach, California, where the Wilmington field has yielded oil from reservoirs at depths of 600–1200 m. The subsidence bowl has deepened to 9 m, and is centred on the industrialized harbour area, where it has caused extensive structural damage and left large areas below sea level; remedial costs had run to over $100 million by 1962 (Mayuga and Allen, 1970). Monitoring via the oil wells showed that nearly

Figure 8.6 Part of Long Beach harbour complex, California, which has suffered severe subsidence above the Wilmington oilfield. The land areas, punctured by numerous oil wells, now lie well below sea level behind their protective embankments.

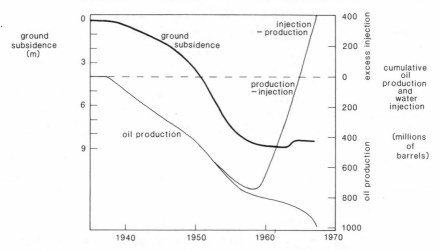

Figure 8.7 Correlation of ground subsidence over the Wilmington oilfield, Long Beach, California, with the oil abstraction which started it and the water injection which stopped it (after Lee, 1979).

70% of the subsidence was due to crushing and fracturing of the feldspar grains in the reservoir sandstone, and ultimate subsidence was estimated to reach up to 15 m (Allen and Mayuga, 1970). Because of the scale of the damage, a massive programme of water injection into the depleting reservoir rocks was started in 1958. Due to the high permeability of the compacting reservoir sandstones, this rapidly halted the subsidence (Figure 8.7). As the water injections repressured the reservoir, surface rebound over the centre of the field exceeded 0.3 m and may ultimately reach double this figure.

Side effects of the Wilmington oilfield subsidence were some small earthquakes created by shearing of critically located shale beds (Lee, 1979), and also horizontal surface movements of up to 3 m around the outer zone of the subsidence bowl. Similar horizontal ground strains were induced over the nearby Inglewood oilfield, which developed its own subsidence bowl 1.2 m deep. Unfortunately in this case an existing fault opened up in the marginal tension zone directly beneath the earth dam retaining the Baldwin Hills Reservoir (Castle *et al.*, 1973; Wilson, 1987), and the subsequent dam failure caused extensive flood destruction in a suburban sector of Los Angeles.

The most recent case of oilfield subsidence has affected the Ekofisk field in the Norwegian sector of the North Sea. Most of the North Sea oil reservoirs are in stable sandstones, but at Ekofisk the chalk reservoir rock at a depth of 3000 m has been crushed due to the loss of fluid support. Subsidence started in 1978, five years after the oilfield was developed, and since then has matched the production profile, now reaching a total of 3.5 m. Ultimate subsidence of 6 m was predicted, and consequently in 1986 the entire Ekofisk platform complex was jacked up by this amount, at a total cost of £200 million (Dadson, 1986) to keep it clear of the North Sea waves and allow production to continue.

Abstraction of gas-bearing water has caused subsidence of more than 2.5 m in Niigata, Japan, and in Italy's Po delta; in the former, the subsidence is now reduced by reinjection of the degassed water, and pumping of the Po gasfield has been stopped by legal action. At Groningen in Holland, withdrawal of natural gas and consequent depressuring of the 2900 m deep reservoir sandstone has promoted subsidence which is predicted to reach an ultimate value of around 300 mm (Schoonbeek, 1977; Pottgens, 1986). Though only a small movement, this is still significant to drainage of the sea-level polders in the subsidence bowl.

Subsidence has been recorded at just some of the sites where geothermal resources of hot water and stream are exploited. The Wairakei site in New Zealand has wells to depths of 600–1200 m, and has induced a subsidence bowl 30 km² in area and 1.5 m deep which is displaced 1500 m laterally from the wellfield (Bixley, 1984). Abstraction is from a series of rhyolite pyroclastics, which clearly permit lateral fluid movement at depth; their compaction appears to be due to a fall in water pressure in the deep aquifer, a fall in steam pressure in the upper part of the aquifer, and also perhaps some intrusion of cold water. In contrast to the Wairakei situation, the California Geysers geothermal field creates much less ground movement, and there is no subsidence at Italy's Lardarello field (Narasimhan and Goyal, 1984). The wide contrasts recorded at geothermal sites mean that subsidence prediction can only be made in the short term using models based on the locally available data.

9 Consolidation of clay soils

The high compressibility of most cohesive clay soils mean that they are prone
to significant compaction under an imposed structural load. The direct
consequence of loading is consolidation of the clay, as the water content is
reduced, and that expulsion of water leads to compaction as the bulk volume
decreases. This causes settlement of the structure which imposes the load on
the clay—a widespread phenomenon, prediction and control of which are
among the main objectives of the engineering science of soil mechanics.
Normal building settlement is not within the scope of this volume. Where the
amount of settlement is large it is often referred to as subsidence, though there
is some confusion over the terminology (Prokopovich, 1985b), and the
geological parameters which normally determine these high settlements
warrant attention in this chapter.

Clay consolidation involves a major primary phase of water expulsion and a
smaller secondary phase of restructuring. The scale of both depends on the
geological properties of the clay; young, organic, montmorillonite-rich clays
with no preconsolidation history exhibit the greatest compression. Rates of
consolidation depend largely on the permeability and thickness of the clays,
which determine the rates of drainage.

Structural loading is not the only cause of clay consolidation. On a large
scale, slow consolidation is due to the natural overburden load of accumulat-
ing sediment, and progressively increases with depth (Skempton, 1970).
Sediment profile dating at Venice shows that mean ground subsidence rates
have ranged 0.4–3.0 mm/year over past millennia (Bortolami et al., 1986), but
this figure incorporates both deep compaction and the crustal sag which
cannot be distinguished until the deep rockhead movement is monitored.

Diagenetic changes involving mineral transformations commonly take
place during deep consolidation over geological time, and are largely
responsible for the drained strength increases after primary consolidation is
complete. Abnormal groundwater conditions can promote changes detri-
mental to the soil properties. Through-drainage by water undersaturated with
respect to silica has transformed kaolinite to gibbsite, with substantial volume
loss, in some clayey sands in Alabama, causing building settlement and small

sinkholes (Isphording and Flowers, 1988), and a similar process may lie behind large surface collapses on laterite soils in Australia (Twidale, 1987).

Settlement of structures on clay

Cohesive clay soils exhibit much larger settlements than granular soils, and the movement falls into four components when structural load is applied (Foott and Koutsoftas, 1984):

(i) Immediate undrained settlement, occurring normally within the construction period, is due to lateral displacement of the soil, and is notably high in some very plastic or organic clays.

(ii) Undrained creep is the slow continuation of the initial compression; normally it is very small and is masked by the increasing rate of consolidation.

(iii) Consolidation settlement is due to the squeezing out of the porewater (the primary consolidation) and accounts for the major part of most ground movements on clay.

(iv) Drained creep, also known as secondary consolidation, is caused by the long-term restructuring of the clay under uniform stress; it progressively declines over time.

The factors which determine the amount of settlement are the geology of the clay, both its mineralogy and its diagenetic changes, the preconsolidation history, and the structural load and foundation shape imposed on it. The processes involved are well reviewed elsewhere (Simons, 1975, 1987; Butler, 1975; Petley and Bell, 1978), but reference is always made back to Terzaghi's consolidation theory and his models of one-dimensional compression. Two- and three-dimensional models, incorporating lateral drainage beneath foundations of small areas, are more difficult to apply, and comparisons of results show that the benefits are often limited on a practical scale (Balasubramaniam and Brenner, 1981). Available techniques of settlement prediction are reviewed in the same paper (and by Foott and Koutsoftas, 1984, and Simons, 1987); generally the final settlement can be predicted to within 20%, but time–settlement prediction remains less reliable. One difficulty in prediction arises from the critical importance of the preconsolidation stress, which is not easily determined in the laboratory due to sample disturbance (Simons, 1975), so full-scale field tests and plate-loading tests can therefore be of major benefit.

In addition to the direct effect of structural loading, clay consolidation can be induced by imposed drainage. On a regional scale, this is mostly due to groundwater abstraction (Chapter 8). On a local scale, seasonal variations of water content in clay soils cause shrinkage and heave, creating annual surface oscillations commonly up to 30 mm and locally up to 50 mm in Britain. Vegetation plays a major role in this, with oak and poplar trees especially abstracting large quantities of soil water. The low montmorillonite content in

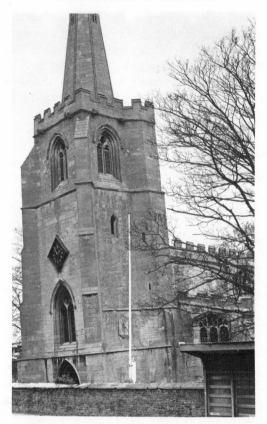

Figure 9.1 Severe settlement of the church tower at Surfleet, on soft alluvial clays in southern Lincolnshire. The tower has only shallow foundations, and has settled differentially as one side is partially supported by the nave of the church.

most of Britain's clays limit the scale of this movement, but the two dry summers of 1975–76 had a significant impact, particularly on the young clays of the London basin (Driscoll, 1983).

The level of acceptable settlement varies considerably with the nature of the structure and its foundations; while 100 mm is a widely tolerated value, curvature through differential settlement is often more critical, and is normally not acceptable above 1:500 (Skempton and MacDonald, 1956). Structural damage due to settlement is reviewed by Burland and Wroth (1975). Where direct loading would cause excessive consolidation of a compressible soil, the settlement can be reduced to acceptable levels by piled, spread or compensated foundations (Zeevaert, 1972, 1987). There is an extensive literature on foundation design, and the Latino Americana Tower in Mexico City provides a good example of successful design for difficult ground conditions.

To reduce settlement on a weak clay, an alternative to designed reduction of unit structural load is soil improvement. Induced soil compression by preloading or surcharging before the main construction phase is accelerated by improving the vertical release of pore water through installed sand or band drains. The benefits of these techniques are reviewed by Pilot (1981) along with dynamic compaction, vibrocompaction and the forming of stone columns (Gambin, 1987). Less widely-used methods include lime stabilization and other forms of chemical treatment (Ingles, 1987).

Severe settlement on clay

High compressibility of clays is largely due to the presence of either organic material or montmorillonite. Included plant debris permits an organic clay to hold an unusually high water content and raises its plasticity; higher contents of plant material give the soil properties which grade towards those of peat (Chapter 10). Settlement characteristics of some highly plastic and organic clays are described by Foott and Ladd (1981), including the case of the Mississippi levées where an added 3 m of fill caused 0.9 m of settlement.

Montmorillonite (the dominant member of the smectite clay mineral group) has a uniquely high capacity for water absorption. For the sodium montmorillonites this can exceed 1500%, orders of magnitude greater than that of the illite and kaolinite clay minerals, while the calcium montmorillonites have intermediate values. Consequently this mineral accounts for most of the excessive compressibility and also the free-swelling capability of clays (Tourtelot, 1974). The worldwide distribution of montmorillonite is restricted by its genesis; it forms most easily from the weathering of volcanic ashes where there is restricted soil leaching in areas of warm climates and moderate seasonal rainfall. It is therefore only a very minor component of most clays in Britain, but volcanic regions in some of the warmer latitudes can provide very difficult ground conditions on clays rich in the mineral.

The level ground on which the major part of Mexico City is built is an old lake bed, and most of the top 30 m of its soils are montmorillonite clays with water contents up to 400%. Mexico City is probably better known for its groundwater pumping and regional subsidence of up to 9 m (Chapter 8), and perhaps 10% of this areal subsidence is due to the imposed load of the city's buildings (Figueroa Vega, 1984). Severe problems have also been caused by the excessive settlement of individual structures. Shallow foundations have been in general use, and buildings up to five storeys high are mostly just on concrete mats. Consequently, heavier buildings have settled, often by a metre within a few years, and the floors of many of the older buildings have sunk below street level (Zeevaert, 1972). Severe structural damage has occurred largely in three situations. Light buildings have subsided differentially into bowls created by heavier adjacent buildings, and the reverse has occurred next to stable buildings on deep piled foundations, though the latter can be remedied by a sheet pile slip surface between the two. Thirdly some buildings have settled badly where a shallow well has drained the soil beneath.

Figure 9.2 Subsidence damage to an old building in Mexico City. Both it and the newer building on the left have only shallow foundations on the very soft montmorillonite clays which characterize the city. Settlement of the new, taller building, structurally sound on its raft, has created the bowl of subsidence which has almost destroyed the older building (Photo: M. Degg).

Mexico City's most famous case of settlement concerns its Palace of Fine Arts. The massive stone structure stands on a concrete mat 2–3 m thick; this imposes a surface loading of 110 kPa, of which over 40% is due to the weight of the mat (Thornley *et al.*, 1955). Construction started in 1904; four years later, the mat and the beginning of the building had already settled 1.65 m, and by 1950 it was 3 m below street level and still going down by 35 mm/year. An early attempt at perimeter sheet piling to stop lateral soil flow had no effect as the settlement was due to primary consolidation, and a programme of grouting beneath the mat merely increased the load.

Just across the street from the Palace of Fine Arts, the Latino Americana Tower, completed in 1951, provides the contrasting successful design (Figure 9.3). It is 140 m high, and has a compensated foundation down to a pad beneath 13 m of basements; supporting this are end-bearing piles driven to a sand at a depth of 33 m (Zeevaert, 1957). Settlement has been about 250 mm, due to consolidation of the lower clay. This has been matched by the subsidence of the street due to the groundwater pumping which has caused

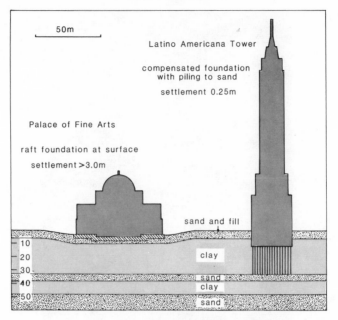

Figure 9.3 The contrast between the foundation design and the resultant settlement of the Palace of Fine Arts and the Latino Americana Tower, both standing on the very soft montmorillonite clays of Mexico City. The drawing is to scale except that the buildings are a little further apart in reality.

compaction of the upper clay; consequently the movable ground floor in the Tower, with individual panels seated on blocks, has not required adjustment. Had the piling been taken through to the lower sand, the Tower would have been stable but would now project above the subsiding streets of the city.

Differential settlement

Natural soils are rarely uniform enough to allow perfectly even settlement. Mexico's Palace of Fine Arts has a distinctive tilt to it, due to differential settlement, but such is rarely noticeable on low, wide structures. Tall, narrow structures show any tilt more obviously, and also exaggerate their tilts when differential loadings are induced by the higher centres of gravity moving away from over the foundation centres.

The classic case of differential settlement is the bell tower of the cathedral at Pisa, which stands 58 m high and now leans 4 m out of true. The Leaning Tower of Pisa has become a major tourist attraction, and is probably the world's most financially successful foundation failure. The Tower stands on a ring of masonry just 19.6 m across founded less than 2 m deep in the soil. It was built between 1174 and 1370, in three spells of about six years each (Figure 9.5), the intervening time gaps being largely due to the disappearances of the

Figure 9.4 The Leaning Tower of Pisa, adjacent to the lower and wider cathedral (Photo: A. Buist).

'architects' when the Tower's movement was already obvious (Spencer, 1953).

The Tower is underlain by sands down to 11 m and then largely by clays to a depth of 40 m (Figure 9.6). Its settlement has been mainly due to compression of the normally consolidated clays between the depths of 11 and 21 m (Mitchell *et al.*, 1977); it has now settled about 0.8 m on the north side and 2.8 m on the south side (Figure 9.5). A modern assessment of the Pisa soils would ascribe to them a safe bearing pressure of around 50 kPa (Spencer, 1953); the original loading was ten times that, and is now nearly 1000 kPa on the south side and almost nil on the north side. The upper sands can bear this, but they are deformed downwards into the compressible clay.

The initial cause of the Tower's lean probably lies in the upper sands and silts, 4–11 m down, which contain more silt and clay to the south (Mitchell *et al.*, 1977, who summarize the data released in 1971 by the Italian Ministry of Public Works). This thesis is supported by cone penetration tests which suggested that, under the load of the Tower, the sands on the north side would compress by 320 mm and those on the south by 400 mm, and also by the rapid

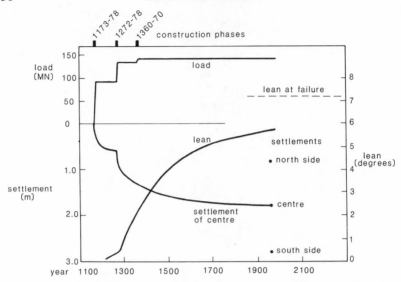

Figure 9.5 The records of loading, settlement and lean of the Tower of Pisa from time of construction to the present (after Mitchell *et al.* 1977). The early data on the lean has been deduced from the modifications made to the masonry as the lean developed during the construction period and the stone courses were corrected to horizontal. The time of ultimate failure cannot be predicted from this graph, as the lean will accelerate in the final stages due to the increasingly eccentric loading.

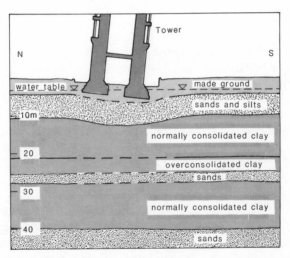

Figure 9.6 Profile through the ground below the Leaning Tower of Pisa, Italy, with the deformed boundary of the sands and clays constructed from borehole data (after Mitchell *et al.* 1977).

early settlements which were characteristic of sand and not clay consolidation. An alternative interpretation (Leonards, 1979) ascribes the initial tilt to local shear failure in the overstressed soft clay, with evidence from some recorded ground heave close to the Tower and also the very variable amount and direction of the earliest movements.

If the Tower had been built in a single phase it would have failed before completion; but the soil strength was increased by the unplanned staged consolidation. The largely consolidated clays beneath the Tower can now bear its load, but drained creep is continuing and the lean is therefore increasing (Figure 9.5). Ultimately the Tower will fall, as the increasing load differential eliminates the benefit of the declining rate of secondary consolidation. Remedial action has been debated for many years but so far without resolution. Modern techniques of underpinning and jacking could stabilize the Tower with piles to the sands 40 m below. Alternative schemes include inclined bereholes to remove some soil beneath the north side (Terracina, 1962), or compression of the soils just north of the Tower with a concrete pad stressed by deep ground anchors (Vedes, 1976).

The initial cause of differential settlement may lie in subtle variations within the soil profile, as at Pisa, but equally can lie with more conspicuous geological features. Contrasts in rockhead depth commonly account for uneven settlement over different thicknesses of compressible soil. Significant rockhead relief may be due to the presence of buried valleys or buried sinkholes (Chapter 3). Some case histories of differential settlement from within Britian have shown the cause to lie in buried valleys, either with natural fills of soft, compressible clay (e.g. Jennings, 1976) or where man-made fills were inadequate for subsequent urban development (e.g. Dearman *et al.*, 1977). The embankments of a multi-level road junction near Plymouth exhibited variable amounts of settlement during their staged construction over both a buried valley with a soft organic clay fill and an area of recently drained marsh (Harris, 1987). Figure 9.7 clearly demonstrates the direct relationship between the larger settlements and the depth to the underlying rockhead.

Where differential settlement produces enough eccentric loading to induce shear failure within the soils, the rotation of the structure can be dramatic. A number of cases are reviewed by McKaig (1962), but the most instructive is that of the Transcona grain elevator in the Canadian Prairies. Built in 1912, this stood 30 m high on a concrete raft 23 × 60 m, which was founded on a stiff upper clay 8 m thick above a few metres of soft or very soft clay. Unfortunately the foundation design was based on the tested strength of the upper clay only. Soon after its first loading with grain, the elevator settled 0.3 m within an hour, and then started to tilt; most of the rotation was achieved inside an hour, and the elevator ended up 27° out of true, leaning against a bank of heaved clay where its one side had subsided 7.2 m (White, 1953). Subsequent analysis (Peck and Bryant, 1953) revealed that the soil failed when the ultimate bearing

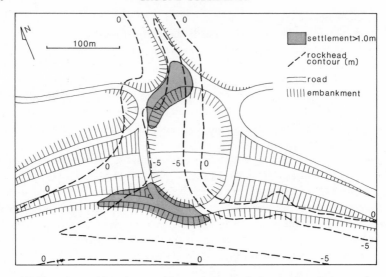

Figure 9.7 Plan of a road junction near Plymouth, England, showing how the areas of greatest settlement of the embankments correlate with the position of the buried valley, roughly bounded by the −5 m rockhead contour (after Harris, 1987).

capacity of the raft, calculated for a weighted mean of the upper and lower clay strengths, was just reached by the full elevator. The initial tilt may have been due to uneven shear in the clays, or perhaps the slightly deeper rockhead and thicker lower soft clay under the side to which it rotated. Subsequently the elevator was emptied, underpinned with bored piles to rockhead, relevelled and returned to use.

10 Subsidence on peat

Peat is unconsolidated soil consisting largely of plant debris in varying degrees of decomposition. It accumulates in poorly-drained areas of upland bog or lowland fen, where anoxic conditions below the water table create an environment of hydrocarbon preservation. The largest areas of peat bog are in the cold, wet, northern latitudes of Canada, Russia and Alaska, while 30% of Finland is surfaced with it. Peat fens are well known in the Netherlands and the Fenlands of England, and are even more extensive in parts of the East Indies and in some coastal zones of the southeastern USA and California.

The high void ratio and extremely low strength of peat create engineering difficulties wherever it occurs. Additionally, peat is prone to regional subsidence when drained; worldwide this occurs at rates of 5–100 mm/year, and has locally accrued to reach total subsidence of over 5 m.

Nature and properties of peat

Peats are highly variable in origin and nature, and have not yet yielded to universal classification. Peat is always defined by its high organic content; muck can be either the same as peat, or an organic soil with a higher mineral content, or a highly decomposed peat. Both peat and muck classify as histosols in the USA. A mire, or a muskeg (the Canadian Indian word) is the system of water, peat and plants whose origins and structure so influence the final form of the peat (Hobbs, 1986).

Environment divides peats into two major groups. Low moor peat accumulates in the lowland fens of deltaic or coastal regions often with warm climates. High moor peats are those of upland bogs generally in colder climates. The low moor can be further subdivided (Stephens *et al.*, 1984) into the sedimentary pond peat, the widespread fibrous peats of sedge and reed marshes which form such rich soils, as in the Florida Everglades, and the woody peats left by climax vegetation in swamp forests. The high moor peat of the northern muskegs and upland blanket bogs is formed mainly of mosses, dominated by the extremely spongy sphagnum.

Unfortunately the peat types cannot be related directly to their engineering strengths, as no peat classification can be quantified. Of critical significance is not only the peat origin but also its degree of humification; this ranges on a scale (von Post, 1922) from 1 to 10, from undecomposed plant debris to a totally humified amorphous sludge, with strength and stability decreasing up the scale.

Combining elements of both humification and plant structure, peats fall into two broad groups—the fibrous (dominated by less humified sedges) and the amorphous-granular (the highly humified sedges and more of the mossy peats). Some of their structures are spectacularly illustrated by Landva and Pheeney (1980). Along with water content, the fibre content is the major guide to the properties of peat, as it adds a natural reinforcement; the more humified, more plastic, less fibrous peats create the main stability problems. The wide range of properties of peats can often be related back to the specific types, and condition, of their plant fibres.

The water content of peat is generally in the range 500–2000% (of dry weight), but is recorded as high as 3235% (MacFarlane, 1969) and is often as low as 100% in peats above the water table. This water may be in macropores, as in the fibrous peats, or in micropores, as in more humified amorphous peats, and the consolidation profile is significantly influenced by the water distribution. Consolidation theory for peat has been developed by Berry and Poskitt (1972); it is complex in its relationship to peat morphology, but has been validated by field tests (Berry and Vickers, 1975). Testing of peat is, however, difficult as laboratory material is always disturbed and unrepresentative, even when collected in large diameter, thin-walled sampling cylinders.

Undrained peat has negligible strength as it acts as a liquid, but drained peat may have an unconfined compressive strength of 20–30 kPa (Hanrahan, 1954). Consolidated by structural load, the strength rises further, and a safe bearing pressure of 70 kPa can be applied on Canadian muskeg (MacFarlane, 1969); however, this value is reduced to 50 kPa on peat over 4.5 m thick, and may be much lower still on some types of peat.

Causes of subsidence

Subsidence on peat may be induced by either loading or drainage. The very high water content of peat ensures that consolidation under structural load is large and also very rapid in the primary stage; the resultant subsidence is on a scale which can be highly destructive unless appropriate design precautions are employed.

Subsidence due to drainage is far more widespread. Where highly porous peat, which accumulated and survived in a saturated state, is left above a declining water table, it suffers volume loss due to shrinkage, consolidation, oxidation and erosion. Shrinkage due to desiccation is rapid and non-reversible, normally accounting for a volume loss of 25–45% (Stephens, 1974; Hobbs, 1986). Consolidation occurs above and below the water table due to loss of groundwater support with a declining head. Elastic compression beneath the water table then causes seasonal subsidence and uplift whose extent depends on both the climate and the efficiency of the land drainage; annual oscillations of 50–100 mm are common (Irwin, 1977; Schothorst, 1977), though Eggelsmann (1986) records up to 200 mm of movement in some undrained reed fen. Consolidation of surface layers is also induced by the tillage which commonly follows wetland drainage. Biochemical oxidation, due to microbial decomposition, above the fallen water table accounts for most of the long-term subsidence of drained peat. It is aided by sporadic burning (either natural or man-induced) and also wind erosion of dry peat on arable farmland. The permanent loss of peat, by oxidation and erosion, is known as wastage, in contrast to the densification induced by shrinkage and consolidation.

Both the rate of subsidence and the relative roles of oxidation and densification vary with drainage depth, climate and peat morphology. Though the processes overlap in their influence, time generally sees the rapid initial subsidence, due to consolidation, eventually overhauled by the ongoing wastage.

Rapid subsidence due to water-table decline

The Fenlands of eastern England are a major region of peat subsidence, with large areas now below sea level even though 50 km from the coast. Originally, the Fens were waterlogged peat, with some floating vegetation, but were drained by various schemes, the main ones starting in the seventeenth century (Darby, 1940). Subsidence was observed soon after each initial drainage, and continued while the land was maintained for agriculture; three pumping stations at successively lower levels had to be built at Prickwillow to lift drainage from the subsiding Burnt Fen into the channel of the River Lark. Each drainage of a virgin Fen produced subsidence of at least 300 mm in the first year, with subsequent movement at declining rates (Astbury, 1958). The only land which did not subside was along the rodhams—meandering strips of silty alluvium which were the low natural levées of the original rivers before drainage was diverted into the new dykes.

Last of the Fens to be drained was that around Whittlesea Mere, which had 400 ha of open water until 1850. A record of the subsequent subsidence was achieved with the Holme Post, a cast iron post standing on timber piles founded in stable clay bedrock 6.7 m below the original peat surface. Through more than 130 years, the Holme Post records the characteristic profile of peat subsidence in response to four phases of pumped drainage (Hutchinson, 1980), and by 1988 the land surface had fallen of total of very nearly 4 m.

The Holme Post record (Figure 10.2) clearly demonstrates that the peat surface always declines rapidly in response to drainage, that the decline is more rapid on the first than on subsequent drawdowns, and that the subsidence rate

Figure 10.1 The Holme Post provides a dramatic record of subsidence on the peat of the English Fenlands. The post is stable as it is founded on bedrock, and its top indicates the ground level 130 years previously, before the peat fen was drained.

then gradually falls off while only wastage continues. This rapid initial subsidence is due to primary consolidation as the groundwater pressure declines, and also to desiccation shrinkage in the upper layers. The subsidence induced by subsequent phases of drainage is then less, due to the deeper layers being already partly consolidated, and compaction of the upper layers through agricultural use is only effective in the first instance.

This pattern of rapid subsidence in response to peat dewatering is repeated elsewhere (Figure 10.3). Large parts of the urban area spreading west from New Orleans are in reclaimed swamp whose peat was 1–4 m thick between new and old channels of the Mississippi (Snowden, 1986). Subsidence in the Kenner area totals nearly 2 m, induced by two phases of early drainage and then the need for more efficient drainage as the New Orleans urbanization spread over the peat. The Florida Everglades are better known for their record of peat wastage but the subsidence profile does show clear response to drainage improvements (Stephens *et al.*, 1984). Subsidence on a relatively

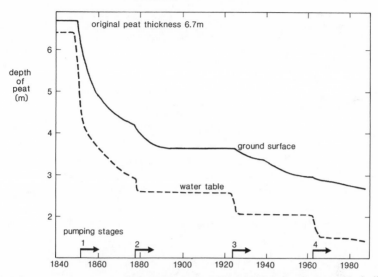

Figure 10.2 The record of peat subsidence through periods of drainage and water-table decline at the Holme Post in the English Fenlands. Pumping capacity has been installed in four stages following an initial gravity drainage of the Fen, and this accounts for some of the irregularities in the wastage profile (after Hutchinson, 1980, with 1988 data added).

small tract of peat in Huntington Beach, California, has reached 5 m (Fairchild *et al.*, 1977), with continued rapid movement in response to unusually complete drainage of the peat due to overdraft in the underlying aquifer. In a contrast of scale, some of the dyked peatlands of Holland have been subsiding for 1000 years with continually maintained shallow drainage, but recent more sudden pumped declines of the water table still promote accelerated subsidence (Schothorst, 1977).

A uniformity in the amount of peat subsidence due to rapid drainage is revealed in Table 10.1. Subsidence of peat drained for the first time is around 60% of the induced groundwater head decline, while the subsidence is only 30% of the head decline where subsequent drainage creates a renewed sudden fall of the water table. These figures, taken from recorded data or interpolated from subsidence profiles extending into wastage, are broadly matched by less well-defined data from peat drainage sites in other countries, and are independent of peat thickness.

Where fen peats in northern Germany are drained to a water table maintained 0.5 m below the surface, Eggelsmann (1986) found rapid subsidence due to primary consolidation on a scale: $s = a(0.080t + 0.066)$, where s is subsidence in metres, t is peat depth in metres, and a is a factor ranging from 1 (for compact peat with $> 12\%$ solids) to 4 (for very loose peat with $< 3\%$ solids). Calculation with this formula does predict subsidence of 55–75% of head loss for thicknesses of typical peat comparable to those in Table 10.1. Correlation of initial subsidence with thickness of peat has also been observed

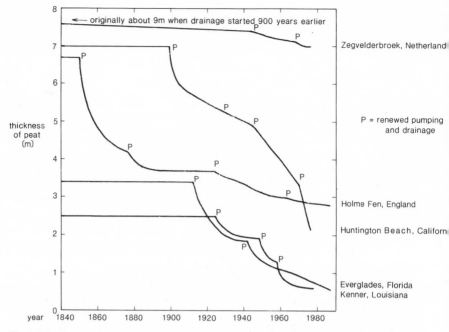

Figure 10.3 Subsidence records of five areas of drained peat, each of which shows direct responses to renewed pumping and dewatering of the peat (after data from Schothorst, 1980; Hutchinson, 1980; Fairchild and Wieber, 1977; Stephens *et al.*, 1984; Snowden, 1986; and 1988 data by author).

Figure 10.4 The concrete post at Belle Glade in the Florida peatland. It is founded on stable bedrock and indicates 63 years of ground subsidence, since the land surface was level with the top of the post before drainage of the peat was initiated.

Table 10.1. Initial subsidence of peat due to rapid drainage*

Peat thickness m	Water-table fall m	Surface subsidence m	WT fall/ subsidence %	Location
				Initial drainage
6.7	2.8	1.8	65	England, Holme Fen
7.0	2.0	1.1	55	California, Huntington Beach
10.0	2.5	1.5	60	England, generalised
				Subsequent drainage
5–8	0.4	0.1	25	Holland, Zegvelderbroek
3.7	0.5	0.15	30	England, Holme Fen

* Data from Fairchild *et al.*, 1977; Hobbs, 1986; Hutchinson, 1980; Schothorst, 1977.

in Louisiana (Snowden *et al.*, 1977) and in the California Delta (Prokopovich, 1985a), but, except in the case of completely drained shallow peat, the water-table decline is more important than thickness in determining peat subsidence.

Based on original work in Poland, Prus-Chacinski (1962) developed a formula which predicts peat subsidence as a function of both drainage and thickness. This can be expressed as: $s = (0.003/G^2) (td^2)^{0.33}$ where d is the water-table decline in metres and G is dry bulk specific gravity of the peat. Restricted to peat less than about 8 m thick, this has been applied successfully at sites in England, and is a useful means of predicting peat subsidence consequent on draining.

Continuing subsidence due to wastage

The Everglades of southern Florida comprise a huge area of peat, originally around 5 m thick, including a 275 000 ha Agricultural Area just south of Lake Okeechobee. Drainage was started in 1906, and the onset of drained aerobic conditions promoted large scale peat destruction and regional subsidence (Stephens, 1958; 1974). At Belle Glade a concrete post 2.7 m long was founded, in 1924, on limestone bedrock beneath peat of the same depth. Frequently photographed, it provides a record of subsidence comparable to that of England's Holme Post, and in 1987 the ground was down to its 3 ft 9 in mark, indicating subsidence of 1.5 m. Continuing subsidence of the Everglades is mostly in the range of 15–30 mm/year (Shih *et al.*, 1979).

Higher subsidence rates are recorded in the California Delta, a 280 000 ha peat area just south of Sacramento (Prokopovich, 1985a). The tule reed peat (described by Weir, 1950), up to 16 m deep, has been drained since it was protected by levées around 1860, subsequently with increased pumping. Current subsidence rates are 57–76 mm/year (Newmarch, 1981), though rates ranging 28–117 mm/year have been recorded, and in some places have accrued to 6.5 m of total subsidence. The only ground not subsiding is on mineral soils from abandoned river channels (Davis, 1963); these now form low meandering ridges similar to the rodhams of the English Fens.

The continuing subsidence of these drained peatlands is due to wastage and secondary consolidation. The latter involves restructuring of the peat under its own load, varies with the fibre content of the peat, and is recognizable by the resultant densification. The rate of consolidation subsidence is related to the depth to the water table, but may also increase on very thick peat and so account for some of the faster subsidence, as in the California Delta. It decreases over time after the initial head decline on drainage, but is independent of climate.

Chemical breakdown of drained peat by microbial oxidation above the water table causes the wastage which is a major mechanism of peatland subsidence. It takes place throughout the depth of the drained peat, with no disturbance of surface vegetation, and the peat is lost as carbon dioxide, whose production rate has been correlated with subsidence in test plots in the Florida Everglades (Stephens, 1958). The major controls on wastage are therefore the depth to the water table and the level of microbial activity as determined by climate.

The influence of water table depth on peat wastage is widely known. Drainage schemes create shallow subsidence troughs along trunk drains which depress the water table (Stephens, 1958; Murashko, 1970), and the reverse occurs where field-margin irrigation ditches leave subsidence bowls in the drier field centres (Shoham and Levin, 1968). For a single site, a linear relationship exists between water-table depth and wastage rate, and Figure 10.5 is constructed with data from controlled experimental peat plots in the Florida Everglades, Indiana and the polders of Holland (Stephens *et al.*, 1984).

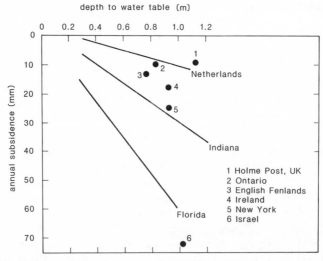

Figure 10.5 Subsidence rates due to peat wastage related to the depth to the water table for regions with contrasting climates (after data from Stephens *et al.*, 1984; Hutchinson 1980; Irwin, 1977; Richardson and Smith, 1977; Armentano, 1980; Levin and Shoham, 1986).

The contrasts between the three line plots in Figure 10.5 are due to the climatic influence, as wastage rises with temperature in response to increased biological oxidation. Overall, the annual wastage of peat varies as 1–7% of water-table depth, depending on climate, and this is confirmed by spot data from other sites also plotted on Figure 10.5. Using a theoretical approach, Stephens and Stewart (1977) quantified biological wastage rates, and field data confirmed that their plot of subsidence against mean annual soil temperature for different water-table depths (Figure 10.6) is representative or slightly conservative. Eggelsmann (1986) used field data to relate wastage to a rainfall factor (N/t; rainfall in mm, over temperature in °C) for normally drained peat with a water-table depth of 55 cm (Figure 10.7); the warmer and drier climates of low rainfall factor showed the most wastage and also the greater contrast between low moor fen and high moor bog, reflecting a botanical influence with less oxidation of woody bog peat. The temperature factor accounts for wastage by oxidation representing around 75% of peat subsidence in Florida, but only 15% of the consequently reduced subsidence of peat in central Russia, the remainder being due to secondary consolidation (Stephens, 1974).

Other factors which influence long-term peat subsidence include the peat morphology, which accounts for some of the scatter of plotted data. Tillage accelerates wastage through turning the soil, and subsidence reduced to zero at Holme Fen (Figure 10.2) through a period of non-arable farming and static water table (Hutchinson, 1980). Subsidence is also reduced by periodic inundation; in the Dutch polders it was less than 2 mm/year when they were flooded each winter, but has accelerated to 6 mm/year since year-round drainage has been maintained (Schothorst, 1977), and subsidence is less in the Bedford Level floodway than on the surrounding drained peat of the

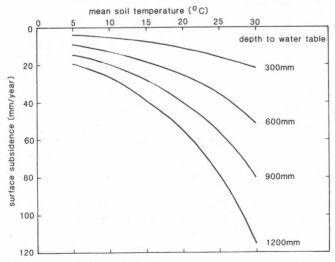

Figure 10.6 Peat wastage rates in terms of mean annual soil temperature and depth to water table (after Stephens and Stewart, 1977).

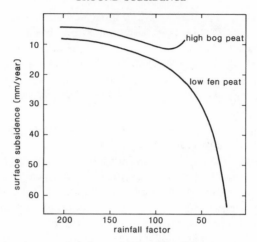

Figure 10.7 Variation in rate of surface subsidence on different types of peat in areas of contrasting climate. The rainfall factor is the millimetres of annual rainfall divided by the mean temperature in °C (after Eggelsmann 1986).

English Fens. Anaerobic decomposition below the water table may be an added factor in parts of the California Delta (Newmarch, 1981). And finally, long series of kept records, such as in the English Fens, show the subsidence rate decreasing over time, except where new drainage is introduced; this is due to less oxidation as less peat remains exposed and also to the reducing rate of consolidation.

Limitation of peat wastage

Drained peat is a wasting asset which can be lost by oxidation 40 times faster than it is formed, and is only farmed at the cost of its ultimate destruction. By the year 2000 nearly half of the Florida Everglades Agricultural Area will have peat soil less than 30 cm thick over barren limestone bedrock (Snyder *et al.*, 1978), and the English Fenlands are already down to less than half their original area of peat soil, though there the peat loss reveals usable clay soils.

These potentially catastrophic soil losses can only be delayed by raising the water table to reduce wastage. Subsidence of the California Delta peatlands could be reduced by 30% by maintaining a higher water table, and further reduced by legislating against the periodic burning, which is of questionable value anyway (Newmarch, 1981). There is a critical need for peatland farming to change to crops which thrive with higher water tables (Snyder *et al.*, 1978); sugar cane therefore becomes unsuitable, and pasture grass or alfalfa should be preferred, while an alternative is a higher water table under the original crop with a small reduction in yield being compensated by the longer life of the land. Other measures are seasonal flooding when the land lies fallow, and growing an aquatic crop such as rice, and both these practices have already been introduced in parts of the Florida Everglades. Wastage may also be reduced by using sprinkler irrigation instead of mole drains, to better wet the upper soil and not just the root zone (Levin and Shoham, 1986). By these means peat subsidence may be controlled and reduced, while unrestrained farm development will eventually destroy the soil it feeds on.

Consolidation of peat

Structural loading of undrained peat induces large settlements, whose amount is proportional to the peat thickness, and whose time is proportional to the square of the thickness (MacFarlane, 1959). Settlement is due to both consolidation and lateral flow, and the latter shears the peat and makes it even weaker. When fill was being tipped for an embankment across the Dandry Mire, for Yorkshire's Settle–Carlisle railway, it just sank into the peat, and adjacent uplift reached 3 m; eventually the bank was abandoned and a low viaduct was founded on bedrock.

Consolidation of peat (reviewed by Berry, 1983) has the primary stage, when free pore water is expelled, generally over in 2–4 days, while the secondary stage, expelling micropore water and loading the peat structure, may last for years, decreasing with the log of time. Samson and La Rochelle (1972) found that shear strength of peat increased from 10 to 90 kPa during consolidation. Settlement increases with imposed load and peat thickness, but also varies with the density and any consolidation history; water content is a useful indicator of potential movement. Stability decreases on less permeable, more humified peats, and increases on coarse fibrous peats (Hobbs, 1986), and also increases with any mineral content. Also, settlement of some Lancashire houses on peat exceeded 200 mm with local variations attributable only to peat's irregular behaviour (Wilson et al., 1985).

The load bearing capacity of peat is low, and Berry (1983) found that houses on rafts exerting loads of only 15 k Pa on 2.5 m of peat still settled 800 mm. Care is needed in extrapolating from laboratory tests, which may predict the amount of settlement but give little idea of time (MacFarlane, 1959). Consolidation tests with 24 h load increments predicted settlement of a Canadian highway which amounted to 40–60% of peat thickness, matched by a fall in water content from 890 to 450% (Samson and La Rochelle, 1972), but Hobbs (1986) emphasizes the unmatched design benefits of large scale field tests for major projects on peat.

Construction on peat

Structural stability on peat may be achieved by either preloading, with or without drainage, excavation or displacement, floating foundations or piling through to bedrock. Techniques are reviewed and defined by MacFarlane (1969) and useful case histories are described by Lea and Brawner (1963) and Samson and La Rochelle (1972).

F

Preloading of peat is usually successful and MacFarlane (1969) outlines a procedure for surcharge design, though in general a surcharge ratio of 1.5–2.0 for a month to a year is adequate on peat 3–9 m thick. In England, surcharge by 2 m of fill for eight months has proved adequate for housing on drained peat (Berry, 1983). A Canadian freeway was built over peat 3–6 m thick, on a sand embankment left for six months and then surcharged with 1.0–1.5 m of sand for a year (Samson and La Rochelle, 1972). Settlement totalled up to 3.3 m, but long-term movements were halved by the surcharge (Figure 10.8). Rebound percentage was found to be five times the surcharge ratio, but it does vary with peat morphology. Hobbs (1986) describes rebound as 2–4% peat thickness, or around 5% of settlement, lasting as long as the surcharge duration, with one third occurring in the first few days, and this matches the Canadian experience.

Sand drains in peat have limited effect, because most peat has high initial permeability and its secondary compression is independent of drainage path length. Restricted vertical permeability may be overcome with wick drains. These were stabbed 17 m into a sequence of clays and peats, on a 1.7 m spacing, to drain into a granular blanket encased in geogrid beneath the Great Yarmouth bypass (Ferguson, 1984). Since then, 8500 wick drains, on a 0.5 m spacing, were used in 15 m of peat under the Ely relief road, and induced rapid settlement of 750 mm.

Mattresses of brushwood or faggots have supported many ancient roads over peat, and the modern version, of bound timber corduroy, is still adequate for minor roads (MacFarlane, 1969). Flexible macadam on thin sub-base also suffices for minor roads in the English Fens, but experimental concrete slab

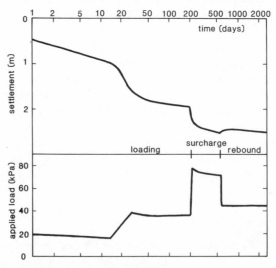

Figure 10.8 Typical profiles of loading and settlement during the construction of a highway across peat near Montreal in Canada. The record extends over five years. covering the two phases of loading, the surcharge stage and the rebound following removal of the surcharge (after Samson and La Rochelle, 1972).

roads have failed badly (Astbury, 1958). Some branch railways float on flexible ballast, and Scotland's Rannoch Moor line is among those which develop a migrating sag as a train crosses the peat bog.

In the English Fenlands, railways over fully drained peat require no special construction, but need a thick ballast blanket over wet peat (Astbury, 1958). The difficulty with any fill or embankment on wet peat is that it may settle more than its thickness, and a lightweight fill is needed to float the structure. Bales of dried peat have been used successfully under Fenland railways and some Irish roads (Hanrahan, 1964), and sawdust supports freeways in Canada (Lea and Brawner, 1963), though these materials only survive where depressed below the water table by an earth cover. Polystyrene blocks and PFA have also been used, the former under the Great Yarmouth bypass where settlement could not be accepted on the embankment approach to a piled bridge (Anon, 1986).

Light buildings can be floated on concrete rafts, preferably with underrims to restrain lateral spreading, but almost always suffer differential settlement. Timber and panelling keep loads low, and central water tanks and chimneys can reduce the subsequent tilt (Astbury, 1958). Better still is the practice of avoiding the peat in the English Fens and building on the rodhams or old islands—as on the Isle of Ely.

Total excavation of peat can be expensive where a dragline is required, as excavators and scrapers often sink in after the surface root mat is taken off; disposal can also be difficult due to excessive spreading of tipped peat.

Figure 10.9 Houses founded on the peat of a drained marsh area close to downtown San Francisco have subsided at virtually the same rate as the ground surface. Regrading of the road has maintained its original level, so that the garages have become inaccessible and new doorways now lead almost directly on to the upper floor.

Figure 10.10 Houses in Belle Glade in the Florida Everglades are founded on piles to the limestone bedrock beneath peat only a few metres thick. Land drainage and the consequent surface subsidence have exposed the foundations, and the front door steps need frequent attention.

Experience with road construction has generally shown removal of peat to be economic only up to thicknesses of 1.5–4.0 m. A displacement technique is better on thicker peat (MacFarlane, 1969). Gravity displacement, where clean sand settles into the peat under its own weight or aided by jetting, is suitable in peat 3–6 m thick; and peat blasting, where a blanket of sand is dropped into the peat which is blasted beneath, works well to depths of 9 m.

Piles driven into or on to rockhead are commonly used for building foundations on peat. An alternative in peat up to 2.5 m deep is a brick-walled cellar, but the cost may not be worth the space benefit. In the New Orleans area friction piles into the underlying clay, to support a concrete slab, are now standard practice for housing and since 1979 are required by law on thick peat in Jefferson county (Snowden, 1986). Roads on piles are expensive, but both the A55 and M4 along the north and south coasts of Wales have piled sections over peat, where a concrete deck was formed on the ground but bears on bored piles; this technique had the added benefit of not disturbing an existing floating railway adjacent to part of the M4.

The long-term problem on peat is that maintained drainage, and hence wastage, continues to lower the ground so that the piled structures end up above ground level—as recorded in nearly all major peat areas. In the Florida Everglades owners of piled houses add a new front door step each 10 years, while Prickwillow Vicarage in the English Fenland now has ten steps up to it. In America soil is often added to the yards (or gardens) to maintain grade, but sealing the void beneath a house creates a trap for gas from subsidence-fractured supply lines; gas is now not installed in the subsiding area of

New Orleans and vents are recommended beneath raised houses (Snowden *et al.*, 1977). A remedy for peat subsidence lies in predraining an area isolated by levées, and then regrading with fill on the subsided peat which is left stable below a reinstated high water table (Kolb and Saucier, 1982), but the cost of this is prohibitive for normal site use. The English Fens lack this precaution, and among the many stone churches, those on piles now stand on mounds raised around them, while those floating on the peat have tilted towers due to the differential loading; the combination of problems is perhaps symbolic of peat subsidence.

11 Hydrocompaction of collapsing soils

Some dry sedimentary materials are prone to internal collapse and consequent volume loss when water is added to them. These are mainly very fine sand and silt soils, either alluvial or loessic, which have remained dry in semi-arid environments. They have porous textures and their strength is derived from intergranular bonds, usually provided by a small percentage of clay; the failure of these bonds when wetted generates the soil collapse.

The process is generally known as hydrocompaction, and the materials susceptible to it are referred to as collapsing soils. It is also sometimes known as 'shallow subsidence', but it is not always surface-related and it can occur to depths of 40 m; the term 'hydroconsolidation' is also sometimes used, but it too is not favoured (Prokopovich, 1986b).

Hydrocompaction can promote ground subsidence of up to 5 m over wide areas. The amount of movement depends on the stress on the soil at the time of wetting. Purely under overburden load, the subsidence normally ranges up to 10% of the thickness of the collapsing soil, and is limited at depth by reduced initial porosity or pre-wetting below the water table. Some soils, known as conditionally collapsing soils (Popescu, 1986), only collapse when wetted under an imposed structural load, and then may exhibit volume losses up to 30%.

Subsidence by hydrocompaction is generally caused by man's activities. These are dominated by land irrigation in arid regions, but also include canal leakage, waste water disposal, other modifications to natural drainage, and even moisture accumulation where soil moisture evaporation is prevented by urbanization. The subsidence can be rapid and destructive, creating ground undulations, fissuring and consequent piping, and severe structural damage, notably to canals, dams, ditches and well casings. On the other hand, gentle water application by irrigation sprinklers causes less damaging subsidence over longer periods.

Potential hydrocompaction of soils may be recognized by saturating samples during laboratory consolidation tests. Figure 11.1 shows two representative test runs, both of which clearly demonstrate the sudden

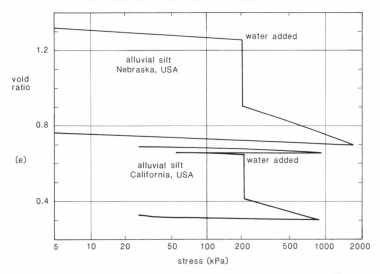

Figure 11.1 Collapse of silt soils when saturated part of the way through consolidation tests; two typical profiles for American collapsing soils (after Dudley, 1970, and Fredrickson, 1977).

decrease in void ratio, and consequent subsidence, due to wetting the soils while under a load of only 200 kPa.

The most widespread collapsible soil is loess, the wind-blown silt characteristic of cold continential interiors. But hydrocompaction is also exhibited by alluvial silts deposited from mudflows, and is locally recorded on some aeolian sands, volcanic ashes and colluvial soils. Though widespread, collapsible soil types can generally be recognized as a type in any one environment and then predicted on a base of local experience (Dudley, 1970). The arid or semi-arid environment is a prerequisite, though the soils may have been deposited by water and subsequently desiccated. The main areas of hydrocompaction in North America are on the loess of the Missouri basin and on the alluvial soils of California (Lofgren, 1969, and a useful later review by Clemence and Finbarr, 1981). In Europe, there is extensive collapsible loess in the Danube basin (Popescu, 1986), but the loess of Western Europe generally has a lower void ratio as it has already collapsed in the modern wetter climate. There are also vast areas of collapsible loess in Russia (see extensive bibliography by Israel Program for Scientific Translations, 1963) and China (Lin and Liang, 1980).

Mechanisms of soil collapse

All collapsing soils are silts or fine sands. Feda (1966) recognized that hydrocompaction only occurs when soil porosity is greater than 40%, and Dudley (1970) confirmed this by finding collapsing soils, with dry densities

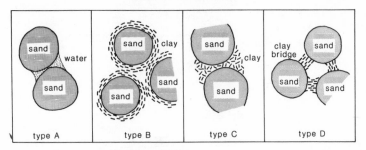

Figure 11.2 Major types of collapsible soil structures, largely created by clay mineral particles forming buttresses and bridges between grains of fine sand and silt. See text for origins of the different types.

ranging from 1.1 to 1.7 t/m^3, which represent porosities of 38–60%. Porosities in excess of 48%, shown by a loose packing with granular contacts, demonstrate the importance of larger voids and bridging clay bonds between the sand or silt grains. Maximum hydrocompaction occurs in soils with 12% clay (Bull, 1964): less clay will not support the metastable bond, and more will compensate the collapse by its own expansion, especially when it includes smectite.

The main types of collapsible soil structures are shown in Figure 11.2 (and photographs of them are to be seen in the useful review by Barden *et al.*, 1973). Type A may be most important where collapse failure is rapid, and a sand–silt mixture merely complicates this structure on a small scale. Authigenic clay, as in some residual soils, mostly creates type B, whereas disordered clay structures of type C tend to form as water is lost by evaporation from alluvial or loessic soils. Prokopovich (1986*b*) suggests this structure may be due to periglacial freezing expansion of the clay followed by effective aeolian desiccation, and has reproduced it by 'freeze-drying' in the laboratory. The bridge structures of type D are formed of clay, in some cases also with calcite or iron oxides, and are mostly in loess.

The hydrocompaction occurs mainly when the addition of water causes dispersion, shearing and failure of the clay bonds. The loss of capillary suction when the soil is saturated promotes the structural collapse, while the low total clay content restricts compensatory expansion. Collapse is most rapid where the suction effect is instantly lost in a type A silt almost devoid of clay, and is normally slower where calcite cements strengthen the clay bonds. Earthquakes are not normally responsible for subsidence on collapsing soils. A block of flats in Ruse, on the Bulgarian side of the Danube, stood on 15 m of loess, and did not move in the 1977 earthquake; but three weeks later it had subsided 250 mm at one end where a fractured water pipe had wetted the loess (Minkov and Evstatiev, 1979). The destructive failures of loess in the same earthquake, and in the many previous quakes in Central China, have been due to landslides.

Hydrocompaction of loess

Loess is widespread through the continental interior mid-latitudes, covering over 15% of the USA, Europe, Russia and China. Most of it is metastable loosely-packed silt with grain size 0.01–0.05 mm which, along with some comparable aeolian sands and volcanic ashes, exhibits hydrocompaction. In arid regions, where it has low moisture contents, it may collapse when wetted, in some cases without any imposed load. New irrigation canals, in the Fergana basin of southern Russia, created severe subsidence, with associated fracturing, fissuring and piping which caused major canal losses (Lofgren, 1969).

Loess is likely to have already collapsed where it has a higher moisture content and greater overburden load at depth, or in regions of wetter climate such as Western Europe. In the huge belt of thick loess in the Hwang He basin of China, only the top, youngest beds are collapsible; rarely is the collapsible material more than 10 m thick, and nowhere does it exceed 20 m (Lin and Liang, 1980).

Above the water table, dry loess may have a safe bearing pressure of 300 kPa or even higher. It is strong enough to have thousands of dwellings cut into its deeper layers in China. But when wetted, hydrocompaction is initiated under imposed loads ranging from 350 kPa down to nil (Lofgren, 1969). Most loess subsidence starts under loads of 70–140 kPa, equivalent to 6–12 m of overburden, hence explaining the dominance of pre-collapsed soil at depth. The extensive loess of the Missouri basin has undisturbed dry densities of 1.17–1.47 t/m^3, and the low density material compacts up to 20% under load when wetted (Figure 11.3). The main hydrocompaction hazard is recognized where the dry density is less than 1.28 t/m^3 and the moisture content is less than 10%, whereas dense loess (over 1.44 t/m^3) or pre-collapsed material with over 15% moisture can support most structures with only minimal settlement (Lofgren, 1969).

Hydrocompaction of alluvial silt

The prime example of hydrocompaction of alluvium is that in the San Joaquin Valley of California, where it is superimposed on even greater subsidence profiles due to groundwater abstraction and clay compaction at depth (Chapter 8). First noted in 1915, in the watered ground around a pumping station, hydrocompaction now affects over 500 km^2 in an area southwest of Fresno. Subsidence is mostly 1–2 m, but locally reaches 5 m.

The collapsible soils in the Valley are silts in the secondary alluvial fans below the smaller valleys from the Coast Ranges. The primary fans are of coarser stream debris where continual fluvial supply never permitted desiccation. Flash floods on the smaller intervening fans created mudflows, which, in the dry climate, were desiccated after deposition and never re-wetted before the next flow buried them. The collapsible silts have a clay content between 5 and

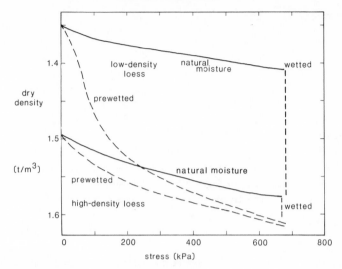

Figure 11.3 Consolidation curves of loess from the Missouri River basin, USA; the lower density soil exhibits over 10% hydrocompaction either when pre-wetted or when wetted after loading at its natural state of low moisture content; the higher density loess suffers only little hydrocompaction (after Clevenger, 1956).

30% (Bull, 1964), and the metastability may be due to the freeze-dried expansion of the clay binder (Prokopovich, 1986*b*). Considerable local variation in the scale of subsidence relates to the patterns of mudflows across the fans (Hall and Carlson, 1965).

A semi-arid climate in the Valley means the soils remain dry, until agricultural irrigation has caused serious hydrocompaction subsidence. Undulating fields, tilted buildings and secondary fissuring are common effects, but the main hazard is to the irrigation canals. Some ditches have been abandoned after five years when they sank below their destination. But some areas are now stable as the soils have been fully wetted by 50 years of induced infiltration.

Previous to major canal construction, hydrocompaction field tests were carried out in the 1950s and 1960s (Lofgren, 1969; Hall and Carlson, 1965). Test plots of 1000 m^2 were flooded behind low dykes, after placement of surface and borehole benchmarks at various depths. Initially the soil had a dry density of 1.1–1.45 t/m^3 and a moisture content of 12–15%, with the water table lying at a depth of 60 m. Infiltration totalled nearly 40 m over 16 months, and subsidence was up to 3.2 m over 27 months. Under no imposed structural load, hydrocompaction started in the surface zones and worked downwards. A collapsible soil with 15 m dry depth subsides almost entirely within a month, but ground above a 60 m thick soil continued to subside for over two years (Figure 11.4). The amount of compaction varied with depth, due to overburden load during and previous to wetting, and reached a maximum of 8.8% at around 25 m depth (Figure 11.5).

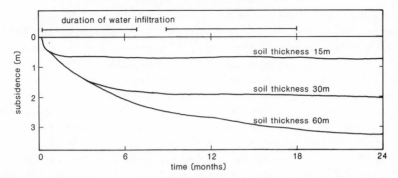

Figure 11.4 Subsidence at a hydrocompaction test site with surface water infiltration in the San Joaquin Valley, California; compaction takes place over longer periods in thicker soils (abstracted from data in Lofgren, 1969).

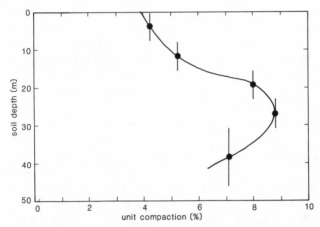

Figure 11.5 Relationship between soil depth and amount of hydrocompaction at a test site in the San Joaquin Valley, California. The vertical bars show the depth range over which the compaction has been measured and meaned (after Lofgren, 1969).

Collapsible alluvia do occur elsewhere. Fractured pipelines have induced hydrocompaction on fine hillwash soils in Johannesburg (De Beer, 1986), and soakaway drains caused subsidence and piping at Luanda airport, built on collapsible ferruginous soil in Angola (Novais-Ferreira and Meireles, 1967). Lofgren (1969) also reported silty alluvium in Washington, with similar properties to that in California; under low loads it hydrocompacted by 6–14%, but with an imposed load of 350 kPa, compaction was up to 20%.

Recognition of potential hydrocompaction

The scale of soil hydrocompaction relates to a number of factors. These include the porosity and moisture content, the clay content and the proportion of smectite, and the imposed load and consolidation history. Various criteria have been proposed to identify collapsing soils. The dry density will be low; Lofgren (1969) recognized the maximum density of a collapsible soil as $1.28 \, t/m^3$ in California, but rising to 1.52 in Colorado, while maxima of 1.6 have been cited by Popescu (1986) and 1.7 by Dudley (1970). As the density alone does not correlate with the silt–clay ratio, it has little predictive value except within a single local environment. Generally the moisture content must be less than 10%, and Lin and Liang (1980) delimited a degree of saturation under 40%. Hydrocompaction is more likely where the clay activity is low, and where the liquid limit is less than 45 and the plasticity index is less than 25 (Dudley, 1970), and is at a maximum where clay content is around 12% (Bull, 1964).

On the premise that a soil, with enough void space to hold the liquid limit water content, will collapse when saturated, Denisov (1951) defined hydrocompaction as likely to occur if $e_L/e_0 < 1$ where e_L is the void ratio at liquid limit and e_0 is initial void ratio. This limit was conservatively redefined by the Soviet Building Code (Lehr, 1967) which stated that hydrocompaction may occur if $(e_0 - e_L)/(1 + e_0) > -0.1$. The original concept may be expressed in terms of liquid limit and dry density (Figure 11.6). Though used with apparent success in early predictions of hydrocompaction in California (Gibbs and Bara, 1967), this criterion was shown, with the benefit of hindsight at the same location, to be unreliable by Prokopovich (1984).

An improved criterion was developed by Feda (1966), working on the Eastern European loess. He defined a subsidence index (K_L) as $K_L = [(W_0/S_0) - W_P]/I_P$ where W_0 is water content, S_0 is degree of saturation, W_P is plastic limit and I_P is plasticity index. Hydrocompaction occurs where the subsidence index is greater than 0.85 and the porosity is over 40%. This concept can be extended (Darwell et al., 1977) to a graphical interpretation which also accounts for dry density and specific gravity.

The commonest and most reliable method of predicting hydrocompaction is the consolidation test; its only disadvantage is the laboratory time involved. A single test run, which includes flooding at either overburden load or at a surcharge load, has proved reliable in regional studies (Fredrickson, 1977; Lin and Liang, 1980). An improvement is the double consolidation test (Figure 11.7), using natural and saturated samples, whose analysis proves more reliable (Jennings and Knight, 1975; Houston et al., 1988). Consolidation tests yield quantitative data related to load, and a collapse potential of over 10% is indicative of severe subsidence, but subsequently measured ground movement frequently exceeds the laboratory determinations.

The alternative to laboratory tests is fieldwork. This can involve simple evaluation based on early indications of subsidence (Prokopovich, 1986), or can cover expensive and time-consuming field flooding tests (as in Lofgren, 1969). Purely under overburden load, subsidence of 5–10% of unit thickness is common in collapsing soils, while imposed loads on shallow soils with high

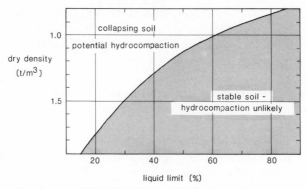

Figure 11.6 A generalized definition of soils which are prone to hydrocompaction, based on the dry density and liquid limit. Mineral density imposes a negligibly small shift of the boundary (after Gibbs and Bara, 1967).

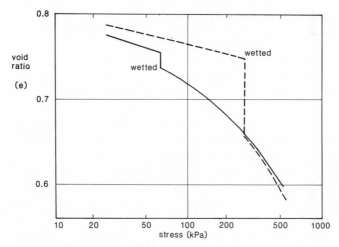

Figure 11.7 Double consolidation test of a pair of samples of a fine sand soil from California, showing collapse and subsequent consolidation when wetted at different stages of loading (after Dudley, 1970).

void ratios may create subsidence of up to 30% over limited thicknesses on first wetting.

Ground treatment of collapsible soils

Hydrocompaction can be prevented by keeping the ground dry, but this is of limited scope and not applicable in irrigation zones. Care with drainage is vital, and heavy structures may still demand floating foundations. Some soils only collapse under both wetting and loading; a Nebraska canal on collapsible loess has remained stable where excavation led to a nil imposed load (Lofgren, 1969).

Where hydrocompaction is predicted, and wetting, perhaps by irrigation, is foreseen, subsidence is best controlled by pre-wetting. Areas of collapsible loam up to 2 m thick have been sprinkle-irrigated and then regraded to render them stable enough for intensive agriculture in Wyoming (Lofgren, 1969). But thicker collapsible soils, notably those over 10 m deep, require more comprehensive treatment.

Most effective is preconstruction flooding, as applied on the San Luis canal projects in California. Sections of the canal route on collapsible soils were pre-wetted by shallow dyked ponds, up to 5 ha in extent, through a two year programme; this created subsidence ranging up to 2.4 m, and was very effective as no canal failures have occurred in the 15 years since they were built (Prokopovich, 1986). For 32 km of canals and 90 km of distributary pipeline, the preflooding cost over $6 million (Prokopovich and Marriott, 1983), and this did not cover additional construction costs to give some stretches of canal extra freeboard to allow for subsequent movement. Subsidence on the deeper soils was effectively accelerated by placing gravel-pack infiltration wells up to 40 m deep on 30 m grids. Elsewhere, controlled hydrocompaction beneath embankments has been successfully promoted during the construction phase by incorporating a basal sand blanket with water fed through perforated pipes (Fredrickson, 1977). Even though preconsolidation is expensive, construction of the San Luis canals would not have been practicable without it.

Alternatives to pre-wetting are limited. Mechanical compaction is less expensive, but is applicable only to thin soils, and does not always stop subsequent hydrocompaction (Prokopovich, 1986b). Russian experience (reviewed by Lofgren, 1969, and Clemence and Finbarr, 1981) has included thermal treatment with hot compressed air or fuel injection, ultrasonic vibration, gaseous silicatization (using carbon dioxide and sodium silicate) and chemical grouting; these techniques are not designed to increase soil strength but just to eliminate potential hydrocompaction, and they have shown some success albeit at high cost.

While hydrocompaction subsidence clearly is controllable, new development on the collapsible soils within the world's drier regions continues to demand relatively expensive ground treatment.

12 Earthquake subsidence

Deformation of the Earth's crust mostly involves large-scale processes whose sheer size places them beyond the control of man. Fortunately they are in the main extremely slow, and subsidence only becomes critical in areas already very close to sea level. The most dramatic and rapid movements are associated with some earthquakes, and various secondary earthquake effects may involve localized ground subsidence. There are also zones of crustal warping which deform by slow, smooth, deep-seated, plastic creep and therefore do not create earthquakes. Strictly these should be referred to as tectonic movements, as they are related to forces within the Earth's interior; most earthquakes are just secondary features of their activity, where sudden stress release creates shock waves and ground vibration.

Tectonic subsidence

Away from the subduction zones, the main cause of tectonic land subsidence is crustal extension and the consequent thinning. Both the London area and the northern part of Holland are currently subsiding by around 2 mm/year due to crustal deformation of the North Sea basin. This subsidence rate is less than the local rise in sea level—a worldwide phenomenon due to the melting of polar ice, but variable in amount due to hydro-isostatic deformation of the Earth's crust. The combination of subsiding land and rising sea level has long-term implications for the coastal defences of Holland, and has also necessitated improved defences around London, including the Thames Barrier.

Tectonically more active areas, such as Japan, may exhibit higher subsidence rates. There, the coastal region around Nagoya is subsiding at 10 mm/year due to crustal warping into the offshore basin. Some of the world's major deltas have formed where sediment accretes in a subsiding basin, and continuing crustal sag accounts for a portion of the land subsidence on these deltas. Tectonic subsidence combined with deep sediment compaction, beneath the Po delta area at Venice is estimated to be about 0.4 mm/year (Carbognin and Gatto, 1986), and a similar or higher rate pertains beneath the Mississippi delta, though at both sites this is overshadowed by sea-level rise, currently about three times that rate, and man-induced subsidence.

Isostatic adjustment causes vertical land movements in response to changes in surface load. Perhaps the most conspicuous has been the uplift following the rapid melting of the thick Pleistocene ice caps; the head of the Gulf of Bothnia, in Scandinavia, is still rising by over 9 mm/year. Comparable subsidence may be due to loading. Large reservoirs are probably the only man-made features heavy enough to induce crustal sag, and Lake Mead, on America's Colorado River, and Lake Kariba, on Africa's Zambezi River, have each created basins subsiding at rates of around 15 mm/year.

Volcanic deflation

More rapid bedrock subsidence may be associated with movements of magma at shallow depths, particularly beneath active volcanoes. Tumescence of volcanic centres, as a result of hot, low-density magma rising inside, is so common that its monitoring is a valuable technique for the prediction of eruptions. Subsidence then occurs during or just after the eruption, due to chamber deflation on loss of support by the extruded magma, and sometimes also at an earlier stage due to sector collapse along grabens. These events are well documented on the Kilauea volcano on Hawaii, especially around the villages on its Kapoho flank, and also on many of the basaltic and andesitic volcanoes of Japan. Widespread subsidence is rarely much more than a metre, and commonly occurs within a time scale of hours; its impact, even where more apparent along a coastline, is usually overshadowed by the associated volcanic damage.

Underground movement of more silicic magmas, with their much higher viscosity (and potential for more explosive eruptions), may last for longer periods and involve greater vertical movement. The classic example is provided by Italy's Phlegrean Fields volcanic complex, now almost overrun by the western suburbs of Naples (and not related to the volcanic cone of Vesuvius, just east of the city). Beneath the cones, craters and caldera of the Phlegrean Fields lies a silicic magma chamber with its roof at a depth of only 3000 m; fluid movement within this has promoted intermittent volcanic activity at the site for 50 000 years, along with the slow vertical oscillations known as bradyseismic activity. These are well recorded at the Roman temple of Serapis, in the modern town of Pozzuoli, where historical data are good and borings by marine animals on the temple columns show the maximum submergence levels. The temple area subsided 12 m in the 1200 years after it was built (Figure 12.2), rose until the nearby volcanic eruption of 1538, and then subsided again. The implications of the renewed inflation between 1969 and 1985 are obvious, and major volcanic destruction of Pozzuoli, and much of Naples, is a serious threat (Berrino et al., 1985). Further round Pozzuoli Bay, the volcanic deflation has also caused long-term subsidence of 14 m at Baia, leaving Roman temples and a port now below sea level (Cotecchia, 1986).

Figure 12.1 The Temple of Serapis at Pozzuoli. Marine borings which reach nearly half way up the two tall columns date from when the temple had subsided below sea level due to volcanic deflation. Photograph taken in 1985 when uplift had left the temple floor nearly dry for the first time in many years.

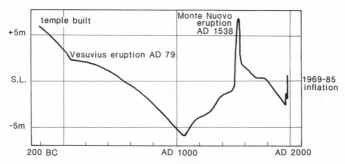

Figure 12.2 The record of historical subsidence and uplift of the Temple of Serapis, at Pozzuoli in southern Italy, which is due to movements of volcanic magma only about 3000 m beneath the site.

Earthquake displacement

The recent ground movements at Pozzuoli have been accompanied by many small but locally destructive earth tremors. Similarly, large earthquakes are the result of intermittent fault movement, usually related to deep-seated plate activity, and overall surface displacement may have a vertical component involving almost instantaneous subsidence of large areas. The famous destructive earthquake in Alaska in 1964 was accompanied by over 100 000 km² of the coastal region subsiding over a metre, while an adjacent region was uplifted. During the 1959 Hebgen Lake earthquake in Montana, 150 km² of hill country subsided more than 3 m. Though careful survey may reveal subsidence after most earthquake events, the impact is rarely of consequence within the context of the widespread surface damage due to ground vibrations.

It is difficult to recognize maximum values for earthquake subsidence. The often-quoted sea-bed subsidence of 210 m of the floor of Sagami Bay, during the great Tokyo earthquake of 1923, was almost certainly greatly modified by sediment shifting and undersea landsliding. In any case, the rarity of events makes useful prediction almost impossible, and the scale and nature of the processes make the ground movements inevitable, and for all practical purposes, unavoidable.

Sediment compaction during earthquakes

During a period of earthquake vibration, an unconsolidated sediment may achieve improved granular packing, resulting in reduction of porosity, loss of volume, and vertical compaction. The consequent surface effects are ground subsidence and water ejection, and where the latter washes out the fine, non-cohesive sediment it creates short-lived 'sand volcanoes', also known as sand blows or sand boils. But it is mainly water that is ejected, so the volume of ground subsidence is far greater than the volume of sand deposition. Earthquake subsidence in this style has affected the coastal sediment plains of South America, around Concepcion in Chile in 1960, and around Chimbote in Peru in 1980, causing permanent flooding in some urban areas. The same mechanism probably accounts for much of the Sunken Lands, flooded in the central Mississippi basin during the 1811–12 earthquakes, though the role of extra crustal displacement is uncertain in this case.

Severe subsidence may also occur, during earthquakes, in the head zones of large landslides. The events at Turnagain Heights and in downtown Anchorage, during the 1964 Alaska earthquake, are well known, but were due to lateral flow and not vertical compaction.

Compaction subsidence is a widely known secondary earthquake phenomenon in areas of unconsolidated sandy sediment with high porosity and where high permeability permits rapid expulsion of the intergranular water. It is just one of the mechanisms that makes structural failure inevitable on areas of unconsolidated soils during earthquakes. Founding on bedrock, or in some cases on soils improved by dynamic vibrocompaction, is a necessary precaution in active seismic regions.

Sediment liquefaction during earthquakes

Far more destructive than compaction is the total liquefaction of some saturated sands during major earthquake events. This occurs when a saturated granular soil, subjected to cyclic loading and reversal of shear stresses, compacts and is unable to drain; the rise in porewater pressure, if maintained,

Figure 12.3 Building subsidence in San Francisco in 1906 due to partial liquefaction of the surface soils during the major earthquake of the same year (photo: G.K. Gilbert, United States Geological Survey).

creates zero effective stress, and hence the soil behaves as a liquid. The process is clearly demonstrated in the laboratory with a tank of sand, a table-tennis ball buried in it, and a steel ball on its surface; when the tank is vibrated, the balls reverse positions. In ground conditions with layered soils, events may be complicated by upward surges of water, expelled from buried liquefied layers, which create delayed liquefaction of the surface soils.

The classic example of liquefaction was at Niigata, Japan, in 1964 (Seed and Idriss, 1967). An earthquake of magnitude 7.3 occurred, with its epicentre 56 km from the city, and after 8 s of ground shaking extensive liquefaction took place. Buildings and off-road vehicles sank into the ground, many by more than a metre, and the four-storey apartment blocks of Kawagishi-cho slowly rotated until one lay on its side, completely intact. At the same time empty storage tanks and sewers floated to the surface. There was also extensive delayed liquefaction, and numerous sand eruptions. Significantly, all the liquefaction occurred in low-lying areas, where the water table was only about a metre below the surface, and the soils were uniformly graded (well-sorted), fine-grained sands up to 30 m thick.

Liquefaction susceptibility must be assessed in respect of the three parameters of soil lithology, groundwater conditions and potential earthquake characteristics. Most liquefiable soils are uniformly graded sands with grain size within the range 0.01–0.5 mm. Sands coarser than 0.7 mm cannot be liquified by a natural earthquake (Seed *et al.*, 1976), and finer sediments are largely liquefaction-resistant cohesive clays. The relative density of the sand is also fundamental, as well-consolidated sands, with high relative density, have little potential for further compaction and pore water expulsion. At Niigata, the benefit of hindsight revealed the importance of relative density, where it was less than 40% at depths of 6 m, in defining the areas of liquefaction failure (Seed and Idriss 1967). In practice this parameter is assessed by the standard penetration test, and the same authors noted SPT N-values over ranges of depths by which liquefaction susceptibility could be predicted. Liquefaction occurred at Niigata where the N-value was still under 20 at depths greater than 10 m.

Sediment lithology may also be viewed in terms of its geological environment, and the highest liquefaction susceptibility was recognized by Youd and Perkins (1978) in very young fluvial and aeolian sands, loess (even of Pleistocene age) and uncompacted artificial fill.

Dry sand, above the water table, will not liquefy beneath level ground through earthquake vibrations, and conversely water tables at the surface offer maximum susceptibility. Where the water table is more than 5 m deep, surface failure is unlikely in any event, as any liquefaction of deep layers would merely dissipate the extruded pore water into a water-table rise (Seed *et al.*, 1976).

The size of the earthquake is clearly critical. The smallest to provide a liquefaction hazard is close to a magnitude 5 event which can be expected to produce peak horizontal acceleration of 0.1 g and cause strong ground shaking up to a kilometre away for less than 10 s. The hazard rises for an event of magnitude 7.5, producing peak acceleration of 0.5 g and ground shaking over 100 km away for perhaps a minute. Prolonged shaking also causes progressive downward liquefaction, and therefore increases delayed surface effects.

Combination of these parameters permits useful prediction of liquefaction susceptibility. Seed *et al.* (1983) define this in terms of the cyclic stress ratio and the normalized N-value (Figure 12.4), though they suggest that the static cone penetration test may be a more useful parameter when a wider data base has been established for it. Compaction caused during the liquefaction depends on the initial relative density, and the resultant ground subsidence has been calculated and compared with observed movements (Tokimatsu and Seed, 1987); it too may be expressed in terms of the N-value and cyclic stress ratio to allow approximate estimations of potential subsidence (Figure 12.4). Predictions were carried into more detail to enable hazard zoning of the Los Angeles area (Tinsley *et al.*, 1985); the liquefaction potential is based on N-values and depths for various water-table conditions and for both major and minor earthquakes (Figure 12.5).

Total avoidance of recognized liquefaction hazard zones is unrealistic in some major conurbations within known seismic belts, and control of earthquakes is not feasible. In areas like Los Angeles or Niigata, liquefaction susceptibility must be recognized and accepted, and the hazard to structures may then be reduced, essentially by either soil densification or drainage (Yoshimi, 1980).

Soil stabilization, notably by dynamic vibrocompaction to increase the

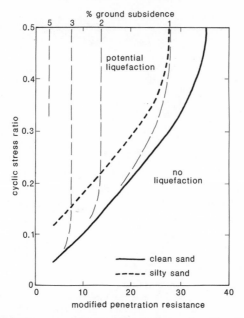

Figure 12.4 The liquefaction potential for sands and silty sands subjected to vibration by an earthquake of magnitude 7.5. The cyclic stress ratio is the ratio of average cyclic shear stress developed horizontally by an earthquake to the initial effective overburden stress on the sand where the latter is close to 4.8 kPa. The modified penetration resistance is the measured value normalized to an effective overburden stress of 4.8 kPa. Mean grain diameter of the sand is > 0.25 mm, and of the silty sand is < 0.15 mm. The potential subsidence is expressed as a percentage of the liquefied sediment thickness, for the sand only (after Seed *et al.*, 1983, and Tokimatsu and Seed, 1987).

relative density, effectively reduces liquefaction susceptibility. Comparable results may be attained by compaction under either a temporary surcharge load or permanent loading. In one area of Niigata, 3 m of artificial fill minimized the damage in 1964, as it increased the overburden stress such that higher shear stresses were required to induce liquefaction and were not reached during the earthquake event (Seed and Idriss, 1967).

Lowering the water table has to be the most effective protective measure, as the shallow unsaturated zone will then not liquefy and the increased effective stress in the undrained zone reduces susceptibility. Seed and Idriss (1967) estimated that if the water table at Niigata had been 1.5 m lower in the 1964 earthquake, there would have been only minimal damage instead of the extensive destruction that occurred. Where water-table lowering by pumping is either impossible or uneconomic, gravel drains, with transmissivity adequate to remove the rapidly expelled pore water, are effective and can also be applied to existing structures. Appropriate drain design is reviewed by Seed and Booker (1977) and Yoshimi (1980), and care has to be taken that water

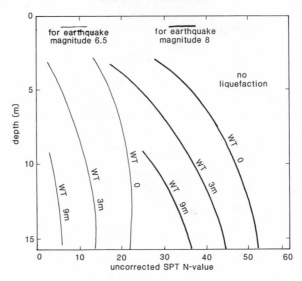

Figure 12.5 Limiting conditions for liquefaction susceptibility during earthquakes of magnitude 6.5 and 8, with assumed peak horizontal accelerations of 0.2 g and 0.5 g respectively. Sediment is assumed to have a mean grain diameter < 0.15 mm, dry unit weight of 1.6 t/m^3 and relative density > 60%. Liquefaction is likely to occur for sediment conditions plotting left of the boundary drawn for the indicated depth to the water table (after Tinsley *et al.*, 1985).

expelled upwards from deeper layers does not threaten liquefaction of the surface layers.

Alternatively, foundation design may protect structures from liquefaction subsidence. Piles to bedrock overcome the hazard, but the behaviour of friction piles is complex during partial liquefaction in layered sediments. Vibratory gravel piles provide generally satisfactory support, and in conjunc-, tion with relief wells are the most widely applied means of liquefaction protection. Light structures, such as wooden houses in Japan, simply float undisturbed when founded on lightweight rafts. Unfortunately, a previous earthquake event does not necessarily provide adequate immunity from liquefaction in a subsequent earthquake. The destruction at Niigata in 1964 followed an earthquake of roughly similar intensity 130 years before, though water tables may have been lower in the earlier event as it preceded widespread subsidence of the coastal plain. Safeguards against liquefaction subsidence will continue to be appropriate in many sandy soil basins within the world's seismic belts.

13 Hazard avoidance

The most effective technique of hazard avoidance is an awareness of the geological environment and the implications that it may hold. Much of the world's land surface has nil potential for ground subsidence, but there are specific geological situations where either slow or catastrophic subsidence can take place. Even though the statistical chances of ground failure may be difficult to quantify in some environments, the accumulated data on these ground conditions, as reviewed in the preceding chapters, leave no excuse for not recognizing a hazard potential.

It is therefore disturbing when a survey of local authorities covering the British coalfields found a high proportion of cases where there was no appreciation of the subsidence hazard presented by old shallow mines (Willis and Chapman, 1980). In areas of redevelopment, the consequence of this can be expensive or disastrous structural failure. Without awareness, site investigation can become a mindless routine instead of a planned evaluation of any difficult ground conditions which could affect the particular site. The site investigation drill which breached a railway tunnel and was then hit by a London Underground train in March 1986 speaks for itself (though this appears to have been due to a survey peg error). In Nottingham, investigation boreholes pierce the roofs of sandstone caves which have walk-in entrances less than 30 m away. These could indicate a lack of planning in the site investigation, and raise the question of how many voids or hazardous sites have not been recognized.

The first stage in subsidence hazard assessment is to identify the significant geological environments. This is normally by recourse to a published geological map, or by reconnaissance geological mapping, as very few subsidence hazards have no surface expression. Figure 13.1 provides a checklist of soils and rocks which could be prone to various styles of subsidence. A list of this nature cannot be exhaustive or definitive, and it does not imply that subsidence will always occur in these ground conditions. It does, however, serve as a reminder that engineering works should only proceed on certain types of ground when the possibilities of subsidence have been fully assessed and any appropriate precautions have been incorporated in the design.

SUBSIDENCE HAZARDS

Rock or soil type	Groundwater abstraction	Drainage infiltration	Any conditions	Old mines	Backfilled quarries	Modern mining	Worst scenario	refer to chapter
Soils								
Peat	■		□				Land drainage	10
Loess and silt		■					Irrigation in arid zones	11
Montmorillonite clay	■						Interbedded sand aquifers	8, 9
Other clays	□			□			Interbedded sand aquifers	8, 9
Clay on limestone	▲	▲	△				Water table falls past rockhead	3
Sand on limestone	■	■	□				Water table falls past rockhead	3
Sand			□		□		Well sorted + earthquake	12
Sedimentary rocks								
Clay	□			□				8
Fireclay				△	□			6
Sandstone				△	□		Building stone	6
Limestone			▲	△	□		Cavernous karst	2, 3
Chalk		△		△	□		Soakaway drains + voids	4
Gypsum		△	□	△				3, 7
Salt	■		□			□	Brining	7
Ironstone				△	□			6
Coal			△	▲	■	■	< 50m deep	5, 6
Other rocks								
Granite				□				6
Basalt			△				Young lavas	2
Volcanoes		□						12
Vein minerals				△	□		Large open stopes	6
Slate				□				6

□ Slow subsidence - mild hazard   △ Collapse potential - limited
■ Slow subsidence - severe hazard   ▲ Collapse potential - severe

Figure 13.1 Checklist of geological environments where ground subsidence may occur and must be considered as a potential hazard.

Once this preliminary assessment has revealed a potential subsidence hazard, the next stage of investigation depends on the nature of the geology. It may be a field search for natural cavities, a laboratory testing programme on soils due to be drained, or a desk study search for data on past mining; the relevant approaches are reviewed in the chapters referred to in Figure 13.1.

A site investigation evolves most economically when it starts with a thorough desk study of available data, and this applies particularly to assessment of both hazards and construction difficulties due to man's past activities (Culshaw and Waltham, 1987). Especially in the developed countries, good geological maps yield much to competent interpretation, and the British

Geological Survey 1:10 000 maps, though covering only part of the country, are of outstanding value. Air photographs are widely available, and Norman and Watson (1975) review the techniques appropriate to identifying sinkholes, peat, gulls, old mines and old shafts, all relevant to subsidence. Records of past mining activity are crucial to subsidence hazard assessment, especially in the coalfields, but the records are seldom complete and therefore always underestimate the surviving threat (Symons, 1978). Exhaustive archive searches run up against the law of diminishing returns, and it is generally necessary to assume all coal seams have been worked until the contrary is proved. Computerized data recording is already offering benefits in hazard assessment, and will improve further when the backlog of archive data is added to the systems.

Engineering geology maps

Subsidence hazards should be recorded on engineering geology maps, and in some areas constitute a significant proportion of the interpretable data. To cover large areas adequately for planning, maps need to be on scales between 1:10 000 and 1:100 000. There are practical limitations to the production and use of such geotechnical maps. They are often too specific in the data they encompass: if wider ranging they still require expert interpretation; they provide useful guidelines, but can rarely offer enough detail for individual engineering works; and they are expensive to produce (Working Party of Geological Society, 1982).

An excellent example of what can be achieved has been produced by the British Geological Survey for a part of central Scotland (Gostelow and Browne, 1986). Separate maps on a scale of 1:50 000 cover: (i) post-glacial sediment thicknesses which range from nil to 70 m above stiff boulder clay or rock; (ii) surface deposits including man-made fill, peat and soft, normally consolidated clays, and (iii) mine workings including all mapped or documented areas of coal mining, areas of suspected shallow mineral workings, known areas of pillar-and-stall mining, and known shafts. A geotechnical planning map is based on a heavy structure bearing pressure of 200 kPa, and divides the area into six zones. The poorest of four zones delimits areas of peat and deep soft clay which do not meet the bearing requirements, and two more unpredictable zones are underlain by known or suspected mine workings below a deep or shallow rockhead. A comparable study of another area in Scotland (Nickless, 1982) includes drift lithology and thickness, opencast areas, shallow undermining and foundation conditions within a suite of 22 thematic maps (mostly concerned with resources), but their preparation on a scale of 1:25 000 limits the detail and potential value. Other British examples of engineering geology mapping, prepared at larger scales, include those for Tyneside (Dearman et al., 1977) and for Nottingham (Figure 13.2); Dearman (1987) reviews the scope of these maps and the need for them to supplement the

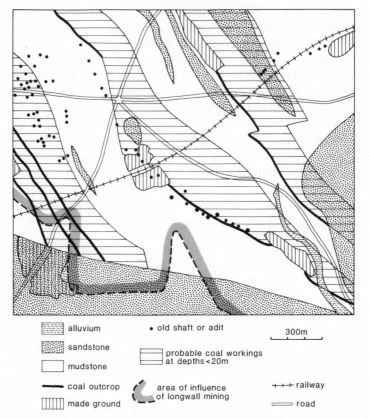

alluvium   • old shaft or adit

sandstone

mudstone   probable coal workings at depths < 20 m

300m

coal outcrop   area of influence of longwall mining   +++ railway

made ground   === road

Figure 13.2 Extract from an engineering geology map of Nottingham (provisional edition, prepared by A. C. Waltham, R. G. Williams, C. Holland and I. Brown), covering part of the western suburbs of the city. Original map compiled at scale of 1:10 000. The area of influence of the longwall mining is delimited by a line 0.7 × depth outside the panels. All the area on this map except the southwest corner has been undermined. There is no current or planned future mining in this area. Many of the shafts near the western margin of the map are shallow bell-pits. Zones with 30 m or 50 m cover above the thick coal seams extend across large portions of the map and are not marked on this extract. Other coal seams do exist, but they are thin, and the likelihood of mines within them is slight. The made ground includes backfilled opencast workings, some probable sites in old quarried areas and tipped material. Only major roads are shown; the eastern half of the mapped area is mostly covered with housing.

conventional geological maps which are not easily read except by professional geologists.

    Geomorphological mapping of surface features has been more widely applied to landslide hazards than to subsidence, and small-scale mapping becomes too generalized for engineering use. Many maps of this type are specific to one parameter, but some excellent multipurpose maps at 1:25 000 have been published for parts of West Germany (Liedtke, 1984). These are full-colour cartographic masterpieces with over 130 symbols (in addition to bedrock geology), of which 25% are related to subsidence processes; but they still require careful interpretation as they are not direct hazard maps.

More detailed mapping, on scales between 1:1000 and 1:10 000 can only be economically prepared on a contract basis preceding major development projects. On this scale geomorphological mapping is more productive. Originally designed to record natural features, it extends to past land use and man-made features, notably due to mining, which may relate to subsidence (Brunsden et al., 1975). Such a map is effectively the documented product of a comprehensive walkover survey, and is a most cost-effective means of land evaluation in the opening stages of a site investigation.

For hazard assessment, small-scale mapping of specific parameters is an alternative to large-scale, multipurpose mapping. Maps of this nature are rarely as widely distributed as conventional maps from national surveys, but many have been published. The liquefaction severity index has been mapped for southern California as a function of probable earthquake vibration amplitude (Chapter 12), and the subsidence hazard can be interpreted when it is superimposed on local sediment maps (Youd and Perkins, 1987). Smaller areas have been covered by liquefaction maps which incorporate the sediment parameters (Tinsley et al., 1985). Hazard zoning of the English chalk (Edmonds et al., 1987) is based on both natural and man-made subsidence factors (Chapter 4).

There are many examples of subsidence hazard zoning in areas of old coal mining. In South Wales, zones delimit rock cover of 6 and 10 times seam thickness over the potentially mined coals (Statham et al., 1987), and comparable results in Tyneside are achieved by zoning on cover depths of 5, 10 and 15 m (Dearman et al., 1977); these maps define the areas most prone to both pillar failure subsidence and crown hole development (Chapter 6). Maps at a 1:10 000 scale divide the Australian mining town of Ipswich into six categories, of which the high risk area is underlain by mine workings less than 40 m deep and is now subject to severe development restrictions (Wood and Renfrey, 1975). The engineering precautions are spelled out in the Ipswich categories, but are available by interpretation from the other maps. There are, however, limits to the amount of data that can be practically represented on a single thematic map, and these impose restrictions on their interpretation. Difficulties commonly arise with regard to areas of known, inferred or suspected undermining, and these factors are usefully reviewed by McMillan and Browne (1987) in the light of experience in the Scottish coalfields. It is important to appreciate that, while thematic maps of this style are valuable for initial planning, they are no substitute for thorough on-site investigation.

Risk assessment

Once the broad potential of subsidence hazard has been recognized for a particular area, it can become significantly more difficult to assess the scale of involved risk. The hazard from rapid collapse events, such as over limestones or old mines, involves probabilities which are not easily evaluated. In contrast, subsidence on peat or soft clay hardly involves probability as it is a certainty, and evaluation of either effect or remedy is purely a balancing of cost and benefit. Each subsidence mechanism involves its own set of parameters and potential remedies, as reviewed in the preceding chapters.

Where the subsidence hazard threatens either property or life, risk assessment involves a number of parameters, only some of which can be quantified. Furthermore, the acceptability of the evaluated risk is subjective, and, especially in public life, involves social factors beyond the scope of this review. There are, however, some legal concepts fundamental to the engineer's responsibilities during planning or construction in an area of potentially hazardous ground conditions. Creation of risk is not illegal, though the engineer must avoid unreasonable risk and must not act carelessly; and risk must be balanced against cost, with the acceptance that total risk elimination may be disproportionately costly beyond a certain level (Cole, 1987a). With

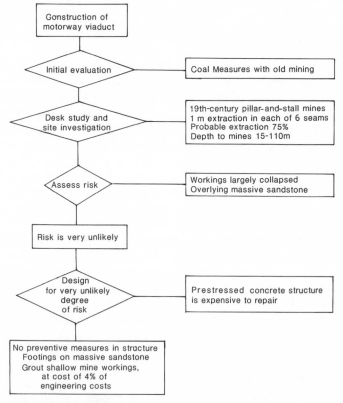

Figure 13.3 An example of the sequence of evaluations and decisions necessary to assess and respond to the risk before construction of a motorway viaduct in an area threatened by subsidence due to the failure of old mineral workings (partly after Cole, 1987).

site investigation results to hand, risk assessment still involves a sequence of subjective evaluations and decisions, where local knowledge and experience are paramount assets. Figure 13.3 takes one example (based partly on Cole, 1987a) where even a very unlikely risk is unacceptable, and grouting of the shallow mine workings was carried out, adding just 4% to the total construction costs.

It is appropriate to note the low cost of precautionary works in the above example. This is normally the case where a subsidence hazard is involved; both thorough site investigation and preventive measures which are incorporated at the design stage rarely add more than a few percent to project costs. Similarly, site relocation can be an economical alternative where a hazard is recognized early by adequate site investigation. Even regional subsidence, through ground dewatering or undermining, need not involve economic risk as the geological processes are known and the costs can be evaluated against the benefits.

Ground subsidence was introduced in Chapter 1 as the poor relation among the family of natural hazards. Perhaps this places it in its true perspective. With few exceptions, subsidence only becomes a real hazard where the geological environment is disturbed by man who forgets or ignores the consequences of his own actions.

References

Akagi, T. (1979) Some land subsidence experiences in Japan and their relevance to subsidence in Bangkok, Thailand. *Geotech. Engg* **10**, 1–48.

Aley, T.J., Williams, J.H. and Massello, J.W. (1972) Groundwater contamination and sinkhole collapse induced by leaky impoundments in soluble rock terrain. Missouri Geological Survey Engineering Geology Ser. **5**, 32pp.

Allen, A.S. (1969) Geologic settings of subsidence. *Geol. Soc. Amer. Rev. Engg Geol.* **2**, 305–342.

Allen, D.R. and Mayuga, M.N. (1970) The mechanics of compaction and rebound, Wilmington oil field, Long Beach, California, USA. International Association Hydrological Sciences Publ. **89**, 410–423.

Andersland, O.B. (1987) Frozen ground engineering. In Bell, F.G. (1987a) *op. cit.*, 8/1–8/24.

Anon (1916) More mine cave-ins threaten parts of Scranton. *Engineering News* **76**, 280–282.

Anon (1986) Great Yarmouth bridge abutment uses polystyrene as lightweight fill. *Ground Engg* **19**(1), 20–23.

Armentano, T.V. (1980) Drainage of organic soils as a factor in the world carbon cycle. *Bio Science* **30**, 825–830.

Arrowsmith, E.J. and Rankilor, P.R. (1981) Dalton By-pass: site investigation in an area of abandoned hematite mine workings. *Q. J. Engg Geol.* **14**, 207–218.

Astbury, A.K. (1958) *The Black Fens.* Golden Head, Cambridge, 217pp.

Balasubramaniam, A.S. and Brenner, R.P. (1981) Consolidation and settlement of soft clay. In Brand, E.W. and Brenner, R.P. (eds.), *Developments in Geotechnical Engineering*, 20: *Soft Clay Engineering*, Elsevier, Amsterdam, 481–566.

Baldwin, M.J. and Newton, M.A. (1987) Basal Permian sand mines and associated surface movements in the Castleford and Pontefract area of West Yorkshire. In Bell, F.G. *et al.* (1987) *op. cit.*, 469–479.

Ballard, R.F., Cuenod, Y. and Jenni, J.P. (1983) Detection of karst cavities by geophysical methods. *Bull. Int. Assn Engg Geol.* **26**, 153–157.

Barden, L., McGown, A. and Collins, K. (1973) The collapse mechanism in partly saturated soil. *Engineering Geol.* **7**, 49–60.

Bates, E.R. (1973) Detection of subsurface cavities. US Army Engineer Waterways Experiment Station Miscellaneous Paper S-73-40, 83pp.

Beck, B.F. (1984) *Sinkholes: their Geology, Engineering and Environmental Impact. Proc. 1st Multidisciplinary Conf. on Sinkholes, Orlando, 1984*, Balkema Rotterdam, 420pp.

Beck, B.F. and Sinclair, W.C. (1986) Sinkholes in Florida: an introduction. Florida Sinkhole Research Institute Report 85-86-4, 16pp.

Beck, B.F. and Wilson, W.L. (eds.) (1987) *Karst Hydrogeology: Engineering and Environmental Applications. Proc. 2nd Multidisciplinary Conf. on Sinkholes and the Environmental Impacts of Karst, Orlando, 1987*, Balkema, Rotterdam, 467pp.

Beggs, T.F. and Ruth, B.E. (1984) Factors affecting the collapse of cavities. In Beck, B.F. (1984) *op. cit.*, 183–188.

Belesky, R.M., Hardy, H.R. and Strouse, F.F. (1987) Sinkholes in airport pavements: engineering implications. In Beck, B.F. and Wilson W.L. (1987) *op. cit.*, 411–417.

Bell, F.G. (ed.) (1975a) *Site Investigations in Areas of Mining Subsidence.* Newnes Butterworths, London, 168pp.

Bell, F.G. (1975b). Salt and subsidence in Cheshire, England. *Engineering Geol.* **9**, 237–247.

Bell, F.G. (1977) A note on the geotechnical properties of chalk. *Engineering Geol.* **11**, 217–225.

Bell, F.G. (1978) Subsidence due to mining operations. In Bell, F.G. (ed.), *Foundation engineering in Difficult Ground*, Butterworth, London, 322–362.

Bell, F.G. (1986) Location of abandoned workings in coal seams. *Bull. Int. Assn Engg Geol.* **33**, 123–132.

Bell, F.G. (ed.) (1987a) *Ground Engineer's Reference Book*. Butterworth, London, 1221pp.

Bell, F.G. (1987b) The influence of subsidence due to present day coal mining on surface development. In Culshaw, M.G. *et al.* (1987) *op. cit.*, 359–367.

Bell, F.G., Culshaw, M.G. and Cripps, J.C. (eds.) (1987) Engineering geology of underground movements. *Proc. 23rd Ann. Conf. of Engineering Group of Geological Society*, Nottingham, 1987 (preprint vol.).

Bell, S.E. (1978) Successful design for mining subsidence. In Geddes, J.D. (1978) *op. cit.*, 562–578.

Benito, G. and Gutierrez, M. (1987) Karst in gypsum and its environmental impact on the Middle Ebro Basin (Spain). In Beck, B.F. and Wilson, W.L. (1987) *op. cit.*, 137–141.

Benson, R.C. and Yuhr, L.B. (1987) Assessment and long term monitoring of localized subsidence using ground penetrating radar. In Beck, B.F. and Wilson, W.L. (1987) *op. cit.*, 161–169.

Bergado, D.T. and Selvanayagam, A.N. (1987) Pile foundation problems in Kuala Lumpur limestone, Malaysia. *Q. J. Engg Geol.* **20**, 159–175.

Bergado, D.T., Nutalaya, P., Balasubramaniam, A.S., Apaipong, W., Chang, C. and Khaw, L.G. (1988) Causes, effects and predictions of land subsidence in AIT campus, Chao Phraya Plain, Bangkok, Thailand. *Bull. Assn Engg Geol.* **25**, 57–81.

Berrino, G., Corrado, G., Luongo, G. and Toro, B. (1985) Ground deformation and gravity changes accompanying the 1982 Pozzuoli uplift. *Bull. Volcanologique* **47**, 187–200.

Berry, P.L. (1983) Application of consolidation theory for peat to the design of a reclamation scheme by preloading. *Q. J. Engg Geol.* **16**, 103–112.

Berry, P.L. and Poskitt, T.J. (1972) The consolidation of peat. *Géotechnique* **22**, 27–52.

Berry, P.L. and Vickers, B. (1975) Consolidation of fibrous peat. *Proc. Amer. Soc. Civil Engrs* **101**, GT8, 741–753.

Bezuidenhout, C.A. and Enslin, J.F. (1970) Surface subsidence and sinkholes in the dolomitic areas of the Far West Rand, Transvaal, Republic of South Africa. *International Association Hydrological Sciences Publ.* **89**, 482–495.

Bixley, P.F. (1984) Case history no. 9.9; the Wairakei geothermal field, New Zealand. In Poland, J.F. (1984) *op. cit.*, 233–240.

Bjelm, L., Follin, S. and Svensson, C. (1983) A radar in geological subsurface investigation. *Bull. Int. Assn Engg Geol.* **26**, 175–179.

Bonaparte, R. and Berg, R.R. (1987) The use of geosynthetics to support roadways over sinkhole prone areas. In Beck, B.F. and Wilson, W.L. (1987) *op. cit.*, 437–445.

Bortolami, G., Carbognin, L. and Gatto, P. (1986) The natural subsidence in the lagoon of Venice, Italy. *International Association Hydrological Sciences Publ.* **151**, 777–784.

Božičević, S. and Pepeonik, Z. (1987) Croatian karst sinkhole experience. In Beck, B.F. and Wilson, W.L. (1987) *op. cit.*, 91–94.

Bozovic, A., Budanur, H., Nonveiller, E. and Pavlin, B. (1981) The Keban Dam foundation on karstified limestone—a case history. *Bull. Int. Assn Engg Geol.* **24**, 45–51.

Brackley, I.J.A., Rosewarne, P.N. and Grady, L.A. (1986) A prediction of surface subsidence caused by lowering the water table in dolomite. *International Association Hydrological Sciences Publ.* **151**, 127–136.

Braithwaite, P.A. and Cole. K.W. (1986) Subsurface investigations of abandoned limestone workings in the West Midlands of England by remote sensors. *Trans. Inst. Ming Metall.* **95A**, 181–190.

Brauner, G. (1973) Calculation of ground movements in European coalfields. *Proc. Fourth Ann. Symp., Illawarra Branch of Australian Institute of Mining and Metallurgy*, 10.1–10.8.

Brink, A.B.A. (1979) *Engineering geology of Southern Africa: Vol. 1.* Building Publications, Pretoria, 319pp.

Brink, A.B.A. (1984) A brief review of the South African sinkhole problem. In Beck, B.F. (1984) *op. cit.*, 123–127.

Bristow, C. (1966) A new graphical resistivity method for detecting air-filled cavities. *Stud. Speleol.* **1**, 204–227.

Brook, G.A. and Allison, T.L. (1986) Fracture mapping and ground subsidence susceptibility modeling in covered karst terrain: the example of Dougherty County, Georgia. International Association Hydrological Sciences Publ. 151, 595–606.

Brown, D.F. and Buist, D.S. (1976) Mine workings and their treatment on the Unstone-Dronfield by-pass. DOE Construction (Department of Environment) 17, 23–25.

Browne, M.A.E., McMillan, A.A. and Forsyth, I.H. (1986) Urban geology: Glasgow's hidden industrial heritage. Geology Today 2, 74–78.

Bruhn, R.W., Magnuson, M.O. and Gray, R.E. (1981) Subsidence over abandoned mines in the Pittsburgh coalfield. In Geddes, J.S. (1981) op. cit., 142–156.

Bruhn, R.W., Speck, R.C. and Thill, R.E. (1983) The Appalachian field: damage to structure above active underground mines. In Sendlein, S.V.A. et al. (1983) op. cit., 657–669.

Brunsden, D., Doornkamp, J.C., Fookes, P.G., Jones D.K.C. and Kelly, J.M.H. (1975) Large scale geomorphological mapping and highway engineering design. Q. J. Engg Geol. 8, 227–253.

Bryan, A. and Bryan, J.G. (1964) Some problems of strata control and support in pillar workings. Ming Eng. London 123, 238–254.

Buist, D.S. and Jones, P.F. (1978) Potential instability of Permian strata in the Pleasley by-pass area, Derbyshire. In Geddes, J.D. (1978) op. cit., 427–448.

Bull, W.B. (1964) Alluvial fans and near-surface subsidence in Western Fresno County, California. US Geological Survey Prof. Pap. 437-A, 71pp.

Bull, W.B. (1973) Geologic factors affecting compaction of deposits in a land subsidence area. Bull. Geol. Soc. Amer. 84, 3783–3802.

Bull, W.B. and Poland, J.F. (1975) Land subsidence due to groundwater withdrawal in the Los Banos-Kettleman City area, California, part 3. Interrelations of water level change, change in aquifer system thickness, and subsidence. US Geological Survey Prof. Pap. 437-G, 62 pp.

Burland, J.B., Hancock, R.J. and May, J. (1983) A case history of a foundation problem on soft chalk. Géotechnique 33, 385–395.

Burland, J.B. and Wroth, C.P. (1975) Settlement of buildings and associated damage. In Settlement of Structures, British Geotechnical Society and Pentech, 611–654.

Burton, A.N. and Maton, P.I. (1975) Geophysical methods in site investigation in areas of mining subsidence. In Bell, F.G. (1975) op. cit., 75–102.

Burton, D. (1978) A three dimensional system for the prediction of surface movements due to mining. In Geddes, J.D. (1978) op. cit., 209–228.

Burton, D. (1985) A program in BASIC for the analysis and prediction of ground movement above longwall panels. In Geddes, J.D. (1985) op. cit., 338–353.

Butler, F.G. (1975) Heavily over-consolidated clays. In Settlement of Structures, British Geotechnical Society and Pentech, 531–578.

Calembert, L. (1975) Problèmes de géologie de l'ingénieur en regions karstiques. Bull. Int. Assn Engg Geol. 12, 93–132.

Calvert, A.F. (1915) Salt in Cheshire. Spon, London, 1206pp.

Cameron, D.W.G. (1956) Menace of present day subsidence due to ancient mineral operations. J. R. Inst. Chartered Surv. 19, 159–171.

Canace, R. and Dalton, R. (1984) A geological survey's cooperative approach to analyzing and remedying a sinkhole related disaster in an urban environment. In Beck, B.F. (1984) op. cit., 343–348.

Carbognin, L. (1983) Land subsidence: a worldwide environmental hazard. Nature and Resources (UNESCO) 21(1), 2–12.

Carbognin, L. and Gatto, P. (1986) An overview of the subsidence of Venice. International Association Hydrological Sciences Publ. 151, 321–328.

Carbognin, L., Gatto, P., Mozzi, G., Gambolati, G. and Ricceri, G. (1977) New trend in the subsidence of Venice. International Association Hydrological Sciences Publ. 121, 65–81.

Carpenter, M.C. and Bradley, M.D. (1986) Legal perspectives on subsidence caused by groundwater withdrawal in Texas, California and Arizona, USA. International Association Hydrological Sciences Publ. 151, 817–828.

Carter, P. (1984) Case histories which break the rules. In Forde, M.C. et al. (1984) op. cit., 20–29.

Carter, P., Jarman, D., and Sneddon, M. (1981) Mining subsidence in Bathgate, a town study. In Geddes, J.D. (1981) op. cit., 101–124.

Castle, R.O., Yerkes, R.F. and Youd, T.L. (1973) Ground rupture in the Baldwin Hills—an alternative explanation. Bull. Assn Engg Geol. 10, 21–46.

Chapman, R.M. (1985) The ground water behaviour pattern and foundation stabilisation practice at Hartebeestfontein. *Bull. Int. Assn Engg Geol.* **31**, 39–50.

Charles, J.A. (1984) Settlement of fill. In Attewell, P.B. and Taylor, R.K. (eds.), *Ground Movements and their Effects on Structures*; Surrey University Press, [Blackie], Glasgow and London, 26–45.

Charles, J.A. and Driscoll, R. (1981) A simple in-situ load test for shallow fill. *Ground Engg* **14**(1), 31–36.

Charles, J.A., Naismith, W.A. and Burford, D. (1978) Settlement of backfill at Horsley restored opencast coal mining site. In Geddes, J.D. (1978) *op. cit.* 229–251.

Charman, J.H. and Cooper, C.E. (1987) The Frindsbury area, Rochester: a review of historical data and their implication on subsidence in an urban area. In Culshaw, M.G. *et al.* (1987) *op. cit.*, 115–124.

Chen, C.Y. and Peng, S.S. (1986) Underground coal mining and attendant subsidence control: some history, technology and research. *Ming Engg* **38**, 95–98.

Chen, J., (1986) Percolation and collapse of reservoirs in karst regions. *Carsologica Sinica* **5**(2), 105–111.

Clark, B.E. (1961) Grouting of Fort Campbell theatre building. *Proc. Amer. Soc. Civil Engrs* **87**, SM2, 33–42.

Clemence, S.P. and Finbarr, A.O. (1981) Design considerations for collapsible soils. *Proc. Amer. Soc. Civil Engrs* **107**, GT3, 305–317.

Clevenger, W.A. (1956) Experiences with loess as foundation material. *Proc. Amer. Soc. Civil Engrs* **82**, SM3, 1–26.

Coates, D.R. (1983) Large scale land subsidence. In Gardner, R. and Scoging, H. (eds.), *Mega-Geomorphology*, Oxford University Press, Oxford, 212–233.

Coker, A.E., Marshall, R. and Thomson, N.S. (1969) Application of computer processed multispectral data to the discrimination of land collapse (sinkhole) prone areas in Florida. In *Proc. 6th Int. Symp. on Remote Sensing of Environment, Ann Arbor, Michigan* **1**, 65–77.

Cole, K.W. (1987*a*) Building over abandoned shallow mines—a strategy for the engineering decisions on treatment. *Ground Engg* **20**(4), 14–30.

Cole, K.W. (1987*b*) Abandoned limestone mines in the West Midlands. *Q. J. Engg Geol.* **20**, 193–198.

Cole, K.W., Turner, A.J. and O'Riordan, N.J. (1984) Limestone mine workings in the West Midlands. In Forde, M.C. *et al.* (1984) *op. cit.*, 40–51.

Collins, J.F.N. (ed.) (1971) Salt: a policy for the control of salt extraction in Cheshire. Cheshire County Council, Chester, 56pp.

Cook, A.F. (1984) Cavity surveillance in abandoned limestone workings in the West Midlands. In Forde, M.C. *et al.*, (1984) *op. cit.*, 30–39.

Cooper, A.H. (1986) Subsidence and foundering of strata caused by the dissolution of Permian gypsum in the Ripon and Bedale areas, North Yorkshire. In Harwood, G.M. and Smith, D.B. (eds.), The English Zechstein and related topics, Geological Society Spec. Publ. **22**, 127–139.

Cooper, R.G., Ryder, P.F. and Solman, K.R. (1982) The windypits in Duncombe Park, Helmsley, North Yorkshire. *Trans. Brit. Cave Res. Assn* **9**, 1–14.

Corbett, B.O. (1984) Abandoned mine workings beneath Monklands District General Hospital, Airdrie. In Forde, M.C. *et al.* (1984) *op. cit.*, 52–63.

Costa, J.E. and Baker, V.R. (1981) *Surficial Geology.* Wiley, New York, 498pp.

Cotecchia, V. (1986) Subsidence phenomena due to earthquakes: Italian cases. International Association Hydrological Sciences Publ. **151**, 829–840.

Crawford, N.C. (1986) Karst hydrological problems associated with urban development: ground water contamination, hazardous fumes, sinkhole flooding and sinkhole collapse in the Bowling Green area, Kentucky. Field trip guidebook, Centre for Cave and Karst Studies, Western Kentucky University, 86pp.

Cruden, D.M., Leung, Y.W. and Thomson, S. (1981) A collapse doline in Wood Buffalo National Park, Alberta, Canada. *Bull. Int. Assn Engg Geol.* **24**, 87–90.

Culshaw, M.G., Bell, F.G., Cripps, J.C. and O'Hara, M. (eds.) (1987) Planning and engineering geology. Geological Society Engineering Geology Spec. Publ. **4**, 641pp.

Culshaw, M.G. and Waltham, A.C. (1987) Natural and artificial cavities as ground engineering hazards. *Q. J. Engg Geol.* **20**, 139–150.

Cunningham, R.R. and Sutherland, W. (1984) The settlement of the house at 31 Main Road, Whitletts, Ayr. In Forde, M.C. *et al.* (1984) *op. cit.*, 64–77.

Dadson, J. (1984) Complex underpin revives St. Wilfrid's. *New Civil Engr* **597**, 22–23.

Dadson, J. (1986) Phillips set for Ekofisk rescue. *New Civil Engr* **698**, 14–16.

Dai, E. and Zhu, Y. (1986) Land subsidence in China. International Association Hydrological Sciences Publ. **151**, 405–414.

Darby, H.C. (1940) *The Draining of the Fens*. Cambridge University Press, Cambridge, 290pp.

Darracott, B.W. and Lake, M.I. (1981) An initial appraisal of ground probing radar for site investigation in Britain. *Ground Engg* **14**(3), 14–18.

Darwell, J.L., Denness, B. and Conway, B.W. (1977) Prediction of metastable soil collapse. International Association Hydrological Sciences Publ. **121**, 544–552.

Davies, B.L. and Smith, R. (1978) The influence of coal mining in the maintenance, design and construction of highway bridges and county-owned structures in South Yorkshire. In Geddes. J.D. (1978) *op. cit.*, 545–561.

Davis, G.H. (1963) Formation of ridges through differential subsidence of peatlands of the Sacramento San Joaquin delta, California. US Geological Survey Prof. Pap. **475-C**, 162–165.

Davis, G.H. (1987) Land subsidence and sea level rise on the Atlantic Coastal Plain of the United States. *Envir. Geol. Water Sci.* **10**, 67–80.

Dean, J.W. (1967). Old mine shafts and their hazards. *Ming Eng. London* **126**, 368–376.

Dearman, W.R. (1987) Land evaluation and site assessment: mapping for planning purposes. In Culshaw, M.G. *et al.* (1987) *op. cit.*, 195–201.

Dearman, W.R., Money, M.S., Coffey, J.R., Scott, P. and Wheeler, M. (1977) Engineering geological mapping of the Tyne and Wear conurbation, north-east England. *Q. J. Engg Geol.* **10**, 145–168.

Dearman, W.R., Strachan, A., Roche, D.P. and Vincett, C. (1982) Influence of mining subsidence on pipelines. *Bull. Int. Assn Engg Geol.* **25**, 19–24.

De Beer, J.H. (1986) Geology of Johannesburg, Republic of South Africa. *Bull. Assn Engg Geol.* **23**, 105–137.

De Bruijn, R.G.M. (1983) Some considerations on the factors that influence the formation of solution pipes in chalk rock. *Bull. Int. Assn Engg Geol.* **28**, 141–146.

de Glopper, R.J. (1986) Subsidence in the recently reclaimed Ijsselmeerpolder, Flevoland. International Association Hydrological Sciences Publ. **151**, 487–496.

de Ruiter, H. (1984) Subsidence of Nanisivik Mines concentrator building. *Ming Mag.* **151**(1), 18–25.

Delattre, N. (1985) Les puits naturels du Tournaisis: étude de leur localisation et contribution à l'étude de leur génèse. *Ann. Soc. Géol. Belg.* **108**, 117–123.

DeMille, G., Shouldice, J.R. and Nelson, H.W. (1964) Collapse structures related to evaporites of the Prairie Formation, Saskatchewan. *Bull. Geol. Soc. Amer.* **75**, 307–316.

Denisov, N.Y. (1951) *The Engineering Properties of Loess and Loess-like Soils* (in Russian). Gosstroiizdat, Moscow, 133pp.

Dougherty, P.H. and Perlow, M. (1987) The Macungie sinkhole, Lehigh Valley, Pennsylvania: cause and repair. In Beck, B.F. and Wilson, W.L. (1987) *op. cit.*, 425–435.

Douglas, I. (1985) Geomorphology and urban development in the Manchester area. In Johnson, R.H. (ed.), *The Geomorphology of North-West England*, Manchester University Press, 337-352.

Driscoll, R. (1983) The influence of vegetation on the swelling and shrinking of clay soils in Britain. *Géotechnique* **33**, 93–105.

Dudley, J.H. (1970) Review of collapsing soils. *Proc. Amer. Soc. Civil Engrs* **96**, SM3, 925–947.

DuMontelle, P.B. and Bauer, R.A. (1983) The mid-continent field: general characteristics of surface subsidence. In Sendlein, L.V.A. *et al.* (1983) *op. cit.*, 671–680.

Dunrud, C.R. (1976) Some engineering geologic factors controlling coal mine subsidence in Utah and Colorado. US Geological Survey Prof. Pap. **969**, 1–39.

Dunrud, C.R. (1984) Coal mine subsidence—Western United States. *Geol. Soc. Amer. Rev. Engg Geol.* **6**, 151–194.

Dunrud, C.R. and Osterwald, F.W. (1978) Coal mine subsidence near Sheridan, Wyoming. *Bull. Assn Engg Geol.* **15**, 175–190.

Dyni, R.C. (1986) Subsidence investigations over salt-solution mines, Hutchinson, Kansas. United States Bureau of Mines Information Circular **9083**, 23pp.

Early, K.R. and Dyer, K.R. (1964) The use of a resistivity survey on a foundation site underlain by karst dolomite. *Géotechnique* **14**, 341–348.

Earp, J.R. and Taylor, B.J. (1986) Geology of the country around Chester and Winsford. *Mem. Geol. Surv. GB*, Sheet 109, 119pp.

Edmonds, C.N. (1983) Towards the prediction of subsidence risk upon the chalk outcrop. *Q. J. Engg Geol.* **16**, 261–266.

Edmonds, C.N. (1987) Induced subsurface movements associated with the presence of natural and artificial underground openings in areas underlain by Cretaceous chalk. In Bell, F.G. *et al.* (1987) *op. cit.*, 239–257.

Edmonds, C.N., Green, C.P. and Higginbottom, I.E. (1987) Subsidence hazard prediction for limestone terrains, as applied to the English Cretaceous chalk. In Culshaw, M.G. *et al.* (1987) *op. cit.*, 283–293.

Edmonds, C.N., Kennie, T.J.M. and Rosenbaum, M.S. (1987) The application of airborne remote sensing to the detection of solution features in limestone. In Culshaw, M.G. *et al.* (1987) *op. cit.*, 125–131.

Ege, J.R. (1984) Mechanisms of surface subsidence resulting from solution extraction of salt. *Geol. Soc. Amer. Rev. Engg Geol.* **6**, 203–221.

Eggelsmann, R.F. (1986) Subsidence of peatland caused by drainage, evaporation and oxidation. International Association Hydrological Sciences Publ. **151**, 497–505.

Egger, K. (1983) Geodetic measurement and the unusual behaviour of the Zeuzier arch dam. *Land Mins Surv.* **1**, 15–21.

Evans, R.T. and Hawkins, A.B. (1985) Significance and treatment of old coal workings at Llanelli Hospital, South Wales. In Geddes, J.D. (1985) *op. cit.*, 188–206.

Evans, W.B. (1970) The Triassic salt deposits of north-western England. *Q.J. Geol. Soc. Lond.* **126**, 103–123.

Evans, W.B., Wilson, A.A., Taylor, B.J. and Price, D. (1968) Geology of the country around Macclesfield, Congleton, Crewe and Middlewich. *Mem. Geol. Surv. GB*, Sheet 110, 328pp.

Evrard, H. (1987) Risques liés aux carrières souterraines abandonnées de Normandie. *Bull. Liais. Lab. Ponts et Chaussées* **150**, 96–108.

Fairchild, J.B., Wiebe, K.H. and Montgomery, J.M. (1977) Subsidence of organic soils and salinity barrier design in coastal Orange County, California. International Association Hydrological Sciences Publ. **121**, 334–346.

Feda, J. (1966) Structural stability of subsident loess soil from Praha-Dejvice. *Engg Geol.* **1**, 201–219.

Ferguson, H. (1984) Bypass gets quick squeeze. *New Civil Engr* **574**, 18–19.

Ferguson, H. (1986) Curtain closes on coalmine cavities. *New Civil Engr* **672**, 22–23.

Ferrians, O.J., Kachadoorian, R. and Greene, G.W. (1969) Permafrost and related engineering problems in Alaska. US Geological Survey Prof. Pap. **678**, 37pp.

Figueroa Vega, G.E. (1984) Case history no. 9.8. Mexico. In Poland, J.F. (1984) *op. cit.*, 217–232.

Foose, R.M. (1969) Mine dewatering and recharge in carbonate rocks near Hershey, Pennsylvania. *Geol. Soc. Amer. Engg Geol. Case Histories* **7**, 45–60.

Foose, R.M. and Humphreville, J.A. (1979) Engineering geological approaches to foundations in the karst terrain of the Hershey Valley. *Bull. Assn Engg Geol.* **16**, 355–381.

Foott, R. and Koutsoftas, D. (1984) Settlement of natural ground under static loadings. In Attewell, P.B. and Taylor, R.K. (eds.), *Ground Movements and their Effects on Structures*, Surrey University Press, [Blackie], Glasgow and London, 1–25.

Foott, R. and Ladd, C.C. (1981) Undrained settlement of plastic and organic clays. *Proc. Amer. Soc. Civil Engrs* **107**, GT8, 1079–1094.

Ford, D. (1988) Characteristics of dissolutional cave systems in carbonate rocks. In Janes, N.P. and Choquette, P.W. (eds.), *Paleokarst*, Springer Verlag, Berlin, 25–57.

Forde, M.C., Topping, B.H.V. and Whittington, H.W. (1984) *Mineworkings 84: Proc. Int. Conf. on Construction in Areas of Abandoned Mineworkings, Edinburgh, 1984*; Engineering Technics Press, Edinburgh, 286pp.

Forrest, J.C.M. and Anderson, A. (1984) Stabilisation of shallow limestone workings at Walsall. In Forde, M.C. *et al.* (1984) *op. cit.*, 78–95.

Franks, C.A.M. (1985) Mining subsidence and landslips in the South Wales coalfield. *Proc. Symp. on Landslides in the South Wales Coalfield, Cardiff*, 225–230.

Fredrickson, R.J. (1977) Foundation treatment for small earth dams on subsiding soils. International Association Hydrological Sciences Publ. **121**, 553–566.

Frye, J.C. and Schoff, S.L. (1942) Deep-seated solution in the Meade Basin and vicinity, Kansas and Oklahoma. *Trans. Amer. Geophys. Union* **23**, 35–39.

Gabrysch, R.K. (1970) Land surface subsidence in the Houston–Galveston region, Texas. International Association Hydrological Sciences Publ. **88**, 43–54.

Gabrysch, R.K. (1984) Case history no. 9.12. The Houston-Galveston region, Texas, USA. In Poland, J.F. (1984) *op. cit.*, 253–262.

Gallagher, C.P., Henshaw, A.C., Money, M.S. and Tarling, D.H. (1978) The location of abandoned mine shafts in rural and urban environments. *Bull. Int. Assn Engg Geol.* **18**, 179–185.

Gallavresi, F. and Rodio, G. (1986) Soil upheaving by grouting to safeguard zones affected by significant subsidence problems: its application to Venice as peculiar example. International Association Hydrological Sciences Publ. **151**, 707–715.

Gambin, M.P. (1987) Deep soil improvement. In Bell, F.G. (1987a) *op. cit.*, 36/1–36/21.

Gambolati, G., Gatto, P. and Freeze, R.A. (1974) Mathematical simulation of the subsidence of Venice. *Water Resources Res.* **10**, 563–577.

Garrard, G.F.G. and Taylor R.K. (1987) Collapse mechanisms of shallow coal-mine workings from field measurements. In Bell. F.G. *et al.* (1987) *op. cit.*, 195–214.

Gatto, P. and Carbognin, L. (1981) The lagoon of Venice—natural environmental trend and man-induced modification. *Hydrol. Sci. Bull.* **26**, 379–391.

Geddes, J.D. (1970) Foundations and structures for unstable valley sites. In *Civil Engineering Problems in the South Wales Valleys*, Institution of Civil Engineers, London, 23–29.

Geddes, J.D. (ed.) (1978) *Large Ground Movements and Structures: Proc. Int. Conf., University of Wales Institute of Science and Technology, Cardiff, 1977*, Pentech, London, 1074pp.

Geddes, J.D. (ed.) (1981) *Ground Movements and Structures: Proc. 2nd Int. Conf., University of Wales Institute of Science and Technology, Cardiff, 1980*, Pentech, London, 1080pp.

Geddes, J.D. (ed.) (1985) *Ground Movements and Structures: Proc. 3rd Int. Conf., University of Wales Institute of Science and Technology, Cardiff, 1984*, Pentech, London, 876pp.

Gendzwill, D.J. and Hajnal, A. (1971) Seismic investigation of the Crater Lake collapse structure of southeastern Saskatchewan. *Can. J. Earth Sci.* **8**, 1514–1524.

Gentile, R.J. (1984) Paleocollapse structures: Longview region, Kansas City, Missouri, *Bull. Assn Engg Geol.* **21**, 229–247.

Gentry, D.W. and Abel, J.F. (1978) Surface response to longwall coal mining in mountainous terrain. *Bull. Assn Engg Geol.* **15**, 191–220.

Gibbs, H.J. and Bara, J.P. (1967) Stability problems of collapsing soil. *Proc. Amer. Soc. Civil Engrs* **93**, SM4, 577–594.

Gilboy, A.E. (1987) Ground penetrating radar: its application in the identification of subsidence solution features–a case study in west-central Florida. In Beck, B.F. and Wilson, W.L. (1987) *op. cit.*, 197–203.

Giles, J.R.A. (1987) Identification of former shallow coal mining from aerial photographs: an example from West Yorkshire. In Culshaw, M.G. *et al.* (1987) *op. cit.*, 133–136.

Godwin, C.G. (1984) Mining in the Elland Flags: a forgotten Yorkshire industry. *Rept Geol. Surv. GB* **16**(4), 1–17.

Gordon, M.M. (1987) Sinkhole repair: the bottom line. In Beck, B.F. and Wilson, W.L. (1987) *op. cit.*, 419–424.

Gostelow, T.P. and Browne, M.A.E. (1986) Engineering geology of the upper Forth Estuary. *Rept. Geol. Surv. GB* **16**(8). 56pp.

Gray, R.E. and Bruhn, R.W. (1984) Coal mine subsidence—eastern United States. *Geol. Soc. Amer. Rev. Engg Geol.* **6**, 123–149.

Gray, R.E. and Meyers, J.F. (1970) Mine subsidence and support methods in Pittsburgh area. *Proc. Amer. Soc. Civil Engrs* **96**, SM4, 1267–1287.

Gregory, O. (1982) Defining the problems of disused coal mine shafts. *Chartered Land Min. Surv.* **3**(4), 4–15.

Greenfield, R.J. (1979) Review of geophysical approaches to the detection of karst. *Bull. Assn Engg Geol.* **16**, 393–408.

Greenfield, R.J., Lavin, P.M. and Parizek, R.R. (1977) Geophysical methods for location of voids and caves. International Association Hydrological Sciences Publ. **121**, 465–484.

Grosch, J.J., Touma, F.T. and Richards, D.P. (1987) Solution cavities in the limestone of Eastern Saudi Arabia. In Beck, B.F. and Wilson, W.L. (1987) *op. cit.*, 73–78.

Guyst, C.A. (1984) Collapse and compaction of sinkholes by dynamic compaction. In Beck, B.F. (1984) *op. cit.*, 419–423.

Hall, C.E. and Carlson, J.W. (1965) Stabilization of soils subject to hydrocompaction. *Bull. Assn Engg Geol.* **2**, 47–58.

Hanrahan, E.T. (1954) An investigation of some physical properties of peat. *Géotechnique* **4**, 108–123.

Hanrahan, E.T. (1954) A road failure on peat. *Géotechnique* **14**, 185–202.

Harris, J. (1987) The design and construction of a grade-separated interchange over soft alluvial and estuarine soils, A38 trunk road, Plymouth. *Q. J. Engg Geol.* **20**, 199–220.

Hatheway, A.W. (1968) Subsidence at San Manuel copper mine, Pinal County, Arizona. *Geol. Soc. Amer. Engg Geol. Case Histories* **6**, 65–81.

Hawkins, A.B. and Privett, K.D. (1981) A building site on cambered ground at Radstock, Avon. *Q. J. Engg Geol.* **14**, 151–167.

Healy, P.R. and Head, J.M. (1984) Construction over abandoned mine workings. Construction Industry Research and Information Association Spec. Publ. **32**, 94pp.

Hellewell, E.G. (1988) The influence of faulting on ground movement due to coalmining: the UK and European experience. *Ming Eng. London* **147**, 334–337.

Helm, D.C. (1977) Estimating parameters of compacting fine-grained interbeds within a confined aquifer system by a one-dimensional simulation of field observations. International Association Hydrological Sciences Publ. **121**, 145–156.

Helm, D.C. (1984) Field-based computational techniques for predicting subsidence due to fluid withdrawal. *Geol. Soc. Amer. Rev. Engg Geol.* **6**, 1–22.

Henry, J.F. (1987) The application of compaction grouting to karstic foundation problems. In Beck, B.F. and Wilson, W.L. (1987) *op. cit.*, 447–450.

Higginbottom, I.E. (1966) The engineering geology of chalk. *Proc. Symp. on Chalk in Earthworks and Foundations*. Institution of Civil Engineers, London, 1–13.

Higginbottom, I.E. (1984) Methods of development in areas of ancient shallow coal mining. In Forde, M.C. *et al.* (1984) *op. cit.*, 273–286.

Hislam, J.L. (1984) Site improvement techniques—mine infilling. In Forde, M.C. *et al.* (1984) *op. cit.*, 121–130.

Hobbs, N.B. (1986) Mire morphology and the properties and behaviour of some British and foreign peats. *Q. J. Engg Geol.* **19**, 7–80.

Holland, E.G. (1962) Hodbarrow Mines, Cumberland. *Mine and Quarry Engg* **28**, 220–227 and 266–274.

Holzer, T.L. (ed.) (1984) Man-induced land subsidence. *Geol. Soc. Amer. Rev. Engg Geol.* **6**, 221pp.

Holzer, T.L. (1984) Ground failure induced by groundwater withdrawal from unconsolidated sediment. *Geol. Soc. Amer. Rev. Engg Geol.* **6**, 67–105.

Hood, M., Evry, R.T. and Riddle, L.R. (1983) Empirical methods of subsidence prediction—a case study from Illinois. *Int. J. Rock Mech. Ming Sci.* **20**, 153–170.

Houston, S.L., Houston, W.N. and Spadola, D.J. (1988) Prediction of field collapse of soils due to wetting. *J. Geotech. Engg ASCE* **114**, 40–58.

Howell, F.T. and Jenkins, P.L. (1977) Some aspects of the subsidences in the rocksalt districts of Cheshire, England. International Association Hydrological Sciences Publ. **121**, 507–520.

Howell, F.T. and Jenkins, P.L. (1985) Centrifuge modelling of salt subsidence features. In Craig, W.H. (ed.), *Application of Centrifuge Modelling to Geotechnical Design*, Balkema, Rotterdam, 193–202.

Hucka, V.J., Blair, C.K. and Kimball, E.P. (1986) Mine subsidence effects on a pressurised natural gas pipeline. *Ming Engg* **38**, 980–984.

Hughes, R.E. (1981) The use of Ordnance Survey bench marks for the study of large-scale mining subsidence. In Geddes, J.D. (1981) *op. cit.*, 185–205.

Hustrulid, W.A. (1976) A review of coal pillar strength formulae. *Rock Mechanics*, **8**, 115–145.

Hutchinson, J.N. (1980) The record of peat wastage in the East Anglian Fenlands at Holme Post, 1846–1978 A.D., *J. Ecol.* **68**, 229–249.

Ireland, R.L., Poland, J.F. and Riley, F.S. (1984) Land subsidence in the San Joaquin Valley, California, as of 1980. US Geological Survey Prof. Pap. **437-I**, 93pp.

Irwin, R.W. (1977) Subsidence of cultivated organic soil in Ontario. *Proc. Amer. Soc. Civil Engrs* **103**, 1R2, 197–205.

Israel Program for Scientific Translation (1963) *Shallow subsidence (hydrocompaction)*. National Science Foundation, Washington, 46pp.

Isphording, W.C. and Flowers, G.C. (1988) Karst development in coastal plain sands: a new problem in foundation engineering. *Bull. Assn Engg Geol.* **25**, 95–104.

James, A.N. and Kirkpatrick, I.M. (1980) Design of foundations of dams containing soluble rocks and soils. *Q. J. Engg Geol.* **13**, 189–198.

Jammal, S.E. (1984) Maturation of the Winter Park sinkhole. In Beck, B.F. (1984) *op. cit.*, 363–369.

Jammal, S.E. (1986) The Winter Park sinkhole and central Florida sinkhole type subsidence. International Association Hydrological Sciences Publ. **151**, 585–594.

Jennings, J.E. (1966) Building on dolomites in the Transvaal. *Civil Engr S. Afr.* **8**(2), 41–62.

Jennings, J.E. and Knight, K. (1975) A guide to construction on or with materials exhibiting additional settlement due to collapse of grain structure. *Proc. 6th Regional Conf. for Africa on Soil Mechanics and Foundation Engineering*, 99–105.

Jennings, J.N. (1985) *Karst Geomorphology*. Blackwell Scientific, Oxford, 293pp.

Jennings, R.A.J. (1976) The problem below. *Q. J. Engg Geol.* **9**, 103–118.

Johnson, K.S. (1987) Development of the Wink Sink in west Texas due to salt dissolution and collapse. In Beck, B.F. and Wilson, W.L. (1987) *op. cit.*, 127–136.

Jones, C.J.F.P. and Spencer, W.J. (1978) The implications of mining subsidence for modern highway structures. In Geddes, J.D. (1978) *op. cit.*, 515–526.

Jones. J.A.A. (1981) The nature of soil piping: a review of research. Monogr. British Geomorphology Research Group **3**, 301pp.

Kang, Y. (1984) Classification of land collapses in karst regions. *Carsologica Sinica* **3**, (2), 146–155.

Khair, A.W. and Peng, S.S. (1985) Causes and mechanisms of massive pillar failure in a southern West Virginia coal mine. *Ming Engg* **37**, 323–327.

Kirk, K.G. and Snyder, E.R. (1977) A preliminary investigation on seismic techniques used to locate cavities in karst terrains. In Dilamarter, R.R. and Csallany, S.C. (eds.), *Hydrologic Problems in Karst Regions*, Western Kentucky University, 79–91.

Kleywegt, R.J. and Enslin, J.F. (1973) The application of the gravity method to the problem of ground settlement and sinkhole formation in dolomite on the Far West Rand, South Africa. *Proc. Symp. on Sinkhole and Subsidence*, International Association Engineering Geology, Hannover, T301–T301.5.

Knight, F.J. (1971) Geologic problems of urban growth in limestone terrains of Pennsylvania. *Bull. Assn Engg Geol.* **8**, 91–101.

Kolb, C.R. and Saucier, R.T. (1982) Engineering geology of New Orleans. *Geol. Soc. Amer. Rev. Engg Geol.* **5**, 75–93.

Koning, H.L. (Co-ordinator) (1986) A detailed case history, Markerwaard, Netherlands. International Association Hydrological Sciences Publ. **151**, 865–928.

Kreitler, C.W. (1977) Fault control of subsidence, Houston, Texas. *Ground Water* **15**, 203–214.

Kumar, R., Singh. B, and Sinha, K.N. (1973) Subsidence investigations in Indian mines. *Proc. Fourth Ann. Symp. Illawarra Branch of Australian Institute of Mining and Metallurgy*, 6.1–6.7.

Lacey, W.D. and Swain, H.T. (1957) Design for mining subsidence. *Architects' J.* **126**, 557–570 and 631–636.

LaMoreaux, P.E. (1984) Catastrophic subsidence, Shelby County, Alabama. In Beck, B.F. (1984) *op. cit.*, 131–136.

LaMoreaux, P.E. and Newton, J.G. (1986) Catastrophic subsidence: an environmental hazard, Shelby County, Alabama, *Envir. Geol. Water Sci.* **8**, 25–40.

Landes, K.K. (1933) Caverns in loess. *Amer. J. Sci.* **25**, 137–139.

Landva, A.O. and Pheeney, P.E. (1980) Peat fabric and structure. *Can. Geotech. J.* **17**, 416–435.

Larson, M.K. and Pewe, T.L. (1986) Origin of land subsidence and earth fissuring, Northeast Phoenix, Arizona. *Bull. Assn Engg Geol.* **23**, 139–165.

Lea, N.D. and Brawner, C.O. (1963) Highway design and construction over peat deposits in lower British Columbia. Highway Research Board, Washington, Research Record **7**, 33pp.

Lee, A.J. (1966) The effect of faulting on mining subsidence. *Trans. Inst. Ming Engrs* **125**, 735–743.

Lee, K.L. (1979) Subsidence earthquake at a California oil field. In Saxena, S.K. (1979) *op. cit.*, 549–564.

Legget, R.F. (1972) Duisburg harbour lowered by controlled coal mining. *Can. Geotech. J.* **9**, 374–383.

Leggo, P.J. and Leech, C. (1983) Sub-surface investigation for shallow mine workings and cavities by the ground impulse radar technique. *Ground Engg* **16**(1), 20–23.

Lehr, H. (1967) Foundation engineering problems in loess soils. *Proc. 3rd Asian Reg. Conf. on Soil Mechanics and Foundation Engg*, 1/6, 20–24.

Leonards, G.A. (1979) Foundation performance of Tower of Pisa. *Proc. Amer. Soc. Civil Engrs* **105**, GT1, 95–105.

Levin, I. and Shoham, D. (1986) Ground elevation loss in the reclaimed Hula Swamp in Israel during the period 1958 to 1980. International Association Hydrological Sciences Publ. **151**, 479–485.

Lewis, W.A. and Croney, D. (1966) The properties of chalk in relation to road foundations and pavements. *Proc. Symp. on Chalk in Earthworks and Foundations.* Institution of Civil Engineers, London, 27–41.

Liedtke, H. (1984) Geomorphological mapping in the Federal Republic of Germany at scales of 1:25,000 and 1:100,000—a priority program supported by the German Research Foundation. Bochumer Geographische Abhandler, **44**, 67–73.

Lin, Z. and Liang, W. (1980) Distribution and engineering properties of loess and loesslike soils in China. *Bull. Int. Assn Engg Geology* **21**, 112–117.

Littlejohn, G.S. (1979*a*) Surface stability in areas underlain by old coal workings. *Ground Engg* **12**(2), 22–30.

Littlejohn, G.S. (1979*b*) Consolidation of old coal workings. *Ground Engg* **12**(4), 15–21.

Littlejohn, G.S. and Head, J.M. (1984) Specification for the consolidation of old shallow mine workings. In Forde, M.C. *et al.* (1984) *op. cit.,* 131–140.

Lofgren, B.E. (1969) Land subsidence due to the application of water. *Geol. Soc. Amer. Rev. Engg Geol.* **2**, 271–303.

Low, E. and Bramwych, M. (1971) Problems in highway engineering through limestone terrain. *Surveyor* **138**(4131), 22–24.

McCann, D.M., Baria, R., Jackson, P.D., Culshaw, M.G., Green, A.S.P., Suddaby, D.L. and Hallam, J.R. (1982). The use of geophysical methods in the detection of natural cavities, mineshafts and anomolous ground conditions. British Geological Survey Engineering Geology Unit Report **82/5**, 272pp.

McCann, D.M., Jackson, P.D. and Culshaw, M.G. (1987) The use of geophysical surveying methods in the detection of natural cavities and mineshafts. *Q. J. Engg Geol.* **20**, 59–73.

McDowell, P.W. (1975) Detection of clay filled sinkholes in the chalk by geophysical methods *Q. J. Engg Geol.* **8**, 303–310.

MacFarlane, I.C. (1959) A review of the engineering characterstics of peat. *Proc. Amer. Soc. Civil Engrs* **85**, SM1, 21–35.

MacFarlane, I.C. (ed.) (1969) *Muskeg Engineering Handbook.* University of Toronto Press, 297pp.

McKaig, T.H. (1962) *Building Failures.* McGraw-Hill, New York, 262pp.

McMillan, A.A. and Browne, M.A.E. (1987) The use or abuse of thematic mining information maps. In Culshaw, M.G. *et al.* (1987) *op. cit.,* 237–245.

Malkin, A.B. and Wood, J.C. (1972) Subsidence problems in route design and construction. *Q. J. Engg Geol.* **5**, 179–194.

Mangan, C. (1985) Indices karstiques et fondations en terrain carbonate. *Am. Soc. Géol. Belg.* **108**, 99–104.

Marsland, A. and Quarterman, R. (1974) Further development of multipoint magnetic extensometers for use in highly compressible ground. *Géotechnique* **24**, 429–433.

Martin, J.C. and Serdengecti, S. (1984) Subsidence over oil and gas fields. *Geol. Soc. Amer. Rev. Engg Geol.* **6**, 23–34.

Mason, D.D.A. (1984) Find that shaft—a review of techniques and cost-effectiveness in the location of abandoned mineshafts. In Forde, M.C. *et al.* (1984) *op. cit.,* 141–150.

Maxwell, G.M. (1975) Some observations on the limitations of geophysical surveying in locating anomalies from buried cavities associated with mining in Scotland. *Ming Eng. London* **134**, 277–285.

Mayuga, M.N. and Allen, D.R. (1970) Subsidence in the Wilmington oil field, Long Beach, California, USA. International Association Hydrological Sciences Publ. **88**, 66–79.

Meade, R.H. (1967) Petrology of sediments underlying areas of land subsidence in central California. US Geological Survey Prof. Pap. **497-C**, 83pp.

Middleton, J. and Watham, T., (1986) *The Underground Atlas: a Gazeteer of the World's Cave Regions.* Hale, London, and St. Martin's, New York, 239pp.

Minkov, M. and Evstatier, D. (1979) On the seismic behaviour of loess soil foundations. *Proc. 2nd US Nat. Conf. on Earthquake Engineering, Stanford,* 988–996.

Mitchell, J.K., Vivatrat, V. and Lambe, T.W. (1977) Foundation performance of Tower of Pisa. *Proc. Amer. Soc. Civil Engrs* **103**, GT3, 227–249.

Moore, H.L. (1987) Sinkhole development along untreated highway ditchlines in East Tennessee. In Beck, B.F. and Wilson, W.L. (1987) *op. cit.,* 115–119.

Morse, C.F.R. (1967) Some mining problems encountered during the construction of a section of the M6 motorway adjacent to Walsall in Staffordshire. *Chartered Surv.* **100**, 236–243.

Murashko, A.J. 1970. Compression of the peat bogs after draining. International Association Hydrological Sciences Publication **89**, 535–546.

Murray, M.J. 1984. Baillieston Interchange bridges (with particular reference to the design for mining subsidence). In Forde, M.C. *et al.* (1984) *op. cit.*, 151–161.

Myers, P.B. and Perlow M. (1984) Development, occurrence and trigger mechanisms of sinkholes in the carbonate rocks of the Lehigh Valley, eastern Pennsylvania. In Beck, B.F. (1984) *op. cit.*, 111–115.

Mylroie, J.E. and Carew, J.L. (1986) Minimum durations for speleogenesis. *Proc. Ninth Int. Speleological Congr. Barcelona*, 1, 249–251.

Narasimhan, T.N. and Goyal, K.P. (1984) Subsidence due to geothermal fluid withdrawal. *Geol. Soc. Amer. Rev. Engg Geol.* 6, 35–66.

National Coal Board (1975) *Subsidence Engineers' Handbook.* National Coal Board Mining Department, London, 111pp.

National Coal Board (1982) *The Treatment of Disused Mine Shafts and Adits.* National Coal Board Mining Department, London, 88pp.

Neighbors, R.J. and Thompson, R.E. (1986) Subsidence in the Houston-Galveston area of Texas. International Association Hydrological Sciences Publ. 151, 785–793.

Newmarch, G. (1981) Subsidence of organic soils. *California Geol.* 34, 135–141.

Newton, J.G. (1984) Review of induced sinkhole development. In Beck, B.F. (1984) *op. cit.*, 3–9.

Newton, J.G. (1986) Natural and induced sinkhole development in the eastern United States. International Association Hydrological Sciences Publ. 151, 549–564.

Newton, J.G. (1987) Development of sinkholes resulting from man's activities in the eastern United States. US Geological Survey Circular 968, 54pp.

Newton, J.G. and Hyde, L.W. (1971) Sinkhole problem in and near Roberts industrial subdivision, Birmingham, Alabama. Geological Survey of Alabama Circular 68, 42pp.

Nickless, E.F.P. (1982) Environmental geology of the Glenrothes district, Fife region: description of 1:25,000 sheet N020. Institute of Geological Sciences Rept 82/15, 59pp.

Nieto, A.S. and Russell, D.G. (1984) Sinkhole development in Windsor-Detroit solution mines and the role of downward mass transfer in subsidence. *In Situ* 8, 293–327.

Nishida, T., Esaki, T., Aoki, K. and Kameda, N. (1986) Evaluation and prediction of subsidence in old working areas and practical preventive measures against mining damage to new structures. International Association Hydrological Sciences Publ. 151, 717–725.

Nonveiller, E. (1982) Treatment methods for soluble rocks. *Bull. Int. Assn Engg Geol.* 25, 165–169.

Norman, J.W. and Watson, I. (1975) Detection of subsidence conditions by photogeology. *Engg Geol.* 9, 359–381.

Novais-Ferreira, H. and Meireles, J.M.F. (1967) On the drainage of muceque—a collapsing soil. *Proc. 4th Reg. Conf. for Africa on Soil Mechanics and Foundation Engineering, Cape Town,* Balkema, Rotterdam, 1, 151–155.

Nuñez, O. and Escojido, D. (1977) Subsidence in the Bolivar Coast. International Association Hydrological Sciences Publ. 121, 257–266.

Nutalaya, P., Chandra, S. and Balasubramaniam, A.S. (1986) Subsidence of Bangkok Clay due to deep well pumping and its control through artificial recharge. International Association Hydrological Sciences Publ. 151, 727–744.

Oates, N.K. (1981) Wild brine extraction and related subsidence in the Cheshire saltfield. Unpubl. thesis, Department of Geology, University of Aston in Birmingham, 360pp.

Orchard, R.J. (1964) Partial extraction and subsidence. *Ming Eng. London* 123, 417–427.

Ove Arup and Partners (1983) Limestone mines in the West Midlands: the legacy of mines long abandoned. Department of the Environment, London, 24pp.

Owen, T.E. (1983) Detection and mapping of tunnels and caves. In Fitch, A.A. (ed.), *Developments in Geophysical Exploration Methods*, Elsevier, Amsterdam, 5, 161–258.

Palmer, A.N. (1984) Geomorphic interpretation of karst features. In LaFleur, R.G. (ed.), *Groundwater as a Geomorphic Agent*, Allen and Unwin, London, 173–209.

Parkinson, J. (1981) Taking the strain off M1's subsiding bridges. *New Civil Engr* 430, 22.

Partridge, T.C., Harris, G.M. and Diesel, V.A. (1981) Construction upon dolomites of the southwestern Transvaal. *Bull. Int. Assn Engg Geol.* 24, 125–135.

Payne, H.R. (ed.) (1986) Mining subsidence: South Wales desk study: summary of research and description of the mapping technique developed. Department of the Environment/Welsh office, Cardiff, 39pp.

Peck, R.B. and Bryant, F.G. (1953) The bearing capacity failure of the Transcona elevator. *Géotechnique* 3, 201–208.

Penman, A.D.M. and Godwin, E.W. (1978) Settlement of experimental houses on land left by opencart mining at Corby. In *Foundations and Soil Technology*, BRE Building Research Series **3**, The Construction Press, [Longman Group], Harlow 280–286.

Petley, D.J. and Bell, F.G. (1978) Settlement and foundations. In Bell, F.G. (ed.), *Foundation Engineering in Difficult Ground*, Butterworth London, 293–321.

Piggott, R.J. and Eynon, P. (1978) Ground movements arising from the presence of shallow abandoned mine workings. In Geddes, J.D. (1978) *op. cit.*, 749–780.

Pilot, G. (1981) Methods of improving the engineering properties of soft clay. In Brand, E.W. and Brenner, R.P. (eds.), *Developments in Geotechnical Engineering, 20. Soft Clay Engineering*, Elsevier, Amsterdam, 637–696.

Poland, J.F. (ed.) (1984) Guidebook to studies of land subsidence due to groundwater withdrawal. Unesco Studies and Reports in Hydrology **40**, 323pp.

Poland, J.F., Lofgren, B.E., Ireland, R.L. and Pugh, R.G. (1973) Land subsidence in the San Joaquin Valley as of 1972. US Geological Survey Prof. Pap. **437-H**, 78pp.

Popescu, M.E. (1986) A comparison between the behaviour of swelling and collapsing soils. *Engg Geol.* **23**, 145–163.

Pottgens, J.J.E. (1986) Ground movements caused by mining activities in the Netherlands. International Association Hydrological Sciences Publ. **151**, 651–659.

Power, J.P. (1985) Dewatering—avoiding its unwanted side effects. American Society of Civil Engineers, New York, 69pp.

Price, D.G. (1972) Engineering geology in the urban environment. *Q. J. Engg Geol.* **4**, 191–208.

Price, D.G., Malkin, A.B. and Knill, J.L. (1969) Foundations of multi-storey blocks on the Coal Measures with special reference to old mine workings. *Q. J. Engg Geol.* **1**, 271–322.

Priest, A.V. and Orchard, R.J. (1958) Recent subsidence research in the Nottinghamshire and Derbyshire coalfield. *Trans. Inst. Ming Engrs* **117**, 500–512.

Prokopovich, N.P. (1972) Land subsidence and population growth. *Proc. 24th Int. Geol. Congr.* **13**, 44–54.

Prokopovich, N.P. (1984) Validity of density—liquid limit predictions of hydrocompaction. *Bull. Assn Engg Geol.* **21**, 191–205.

Prokopovich, N.P. (1985a) Subsidence of peat in California and Florida *Bull. Assn Engg Geol.* **22**, 395–420.

Prokopovich, N.P. (1985b) Land subsidence terminological confusion. *Bull. Assn Engg Geol.* **22**, 106–108.

Prokopovich, N.P. (1986a) Classification of land subsidence by origin. International Association Hydrological Sciences Publ. **151**, 281–290.

Prokopovich, N.P. (1986b) Origin and treatment of hydrocompaction in the San Joaquin Valley, CA, USA. International Association Hydrological Sciences Publ. **151**, 537–546.

Prokopovich, N.P. (1986c) Economic impact of subsidence on water conveyance in California's San Joaquin Valley, USA. International Association Hydrological Sciences Publ. **151**, 795–804.

Prokopovich, N.P. and Marriott, M.J. (1983) Cost of subsidence to the Central Valley Project, California. *Bull. Assn Engg Geol.* **20**, 325–332.

Prudic, D.E. and Williamson, A.K. (1986) Evaluation of technique for simulating a compacting aquifer system in the Central Valley of California, USA. International Association Hydrological Sciences Publ. **151**, 53–63.

Prus-Chacinski, T.M. (1962) Shrinkage of peat lands due to drainage operations. *J. Inst. Water Engrs* **16**, 436–448.

Pyne, R. and Randon, D.V. (1986) Some environmental aspects of the Selby Coalfield. *Ming Eng. London* **146**, 77–84.

Quinlan, J.F. (1986) Legal aspects of sinkhole development and flooding in karst terranes: 1, review and synthesis. *Envir. Geol. Water Sci.* **8**, 41–61.

Quinlan, J.F., Ewers, R.O., Ray, J.A., Powell, R.L. and Krothe, N.C. (1983) Groundwater hydrology and geomorphology of the Mammoth Cave region, Kentucky, and of the Mitchell Plain, Indiana. In Shaver, R.H. and Sunderman, J.A. (eds.), *Field Trips in Midwestern Geology: Bloomington, Indiana*, Geological Society of America and Indiana Geological Survey, **1**, 1–85.

Raghu, D. (1987) Determination of pile lengths and proofing of the bearing stratum of piles in cavernous carbonate formations. In Beck, B.F. and Wilson, W.L. (1987) *op. cit.*, 397–402.

Rat, M. (ed.) (1977) Détection des cavités souterraines. *Bull. Liais. Lab. des Ponts et Chaussées* **92**, 57–86.

Rau, J.L., Nutalaya, P. and Boonsener, M. (1982) Subsidence and chloride contamination at Nong Bo Reservoir, northeast Thailand. *Geotech. Engg* **13**, 51–72.

Reeves, A. (1984) Legal aspects of development in coal mining areas: the National Coal Board involvement. In Forde, M.C. *et al.* (1984) *op. cit.*, 189–195.

Reitz, H.M. and Eskridge, D.S. (1977) Construction methods which recognize the mechanics of sinkhole development. In Dilamarter, R.R. and Casallany, S.C. (1977) *Hydrologic Problems in Karst Regions*, Western Kentucky University, 432–438.

Richardson, S.J. and Smith, J. (1977) Peat wastage in the East Anglian Fens. *J. Soil Sci.* **28**, 485–489.

Samson, L. and La Rochelle, P. (1972) Design and performance of an expressway constructed over peat by preloading. *Can. Geotech. J.* **9**, 447–466.

Sandy, J.D., Piesold, D.D.A., Fleischer, V.D. and Forbes, P.J. (1976) Failure and subsequent stabilisation of no. 3 dump at Mufulira mine, Zambia. *Trans. Inst. Ming Metall.* **85A**, 144–162.

Saxena, S.K. (ed.), (1979) *Evaluation and Prediction of Subsidence*. American Society of Civil Engineers, New York, 594pp.

Schoonbeek, J.B. (1977) Land subsidence as a result of gas extraction in Groningen, The Netherlands. International Association Hydrological Sciences Publ. **121**, 267–284.

Schothorst, C.J. (1977) Subsidence of low moor peat soils in the Western Netherlands. *Geoderma* **17**, 265–291.

Schroeder, J., Beaupré, M. and Cloutier, M. (1986) Ice-push caves in platform limestones of the Montreal area. *Can. J. Earth Sci.* **23**, 1842–1851.

Seed, H.B. and Booker, J.R. (1977) Stabilisation of potentially liquefiable sand deposits using gravel drains. *Proc. Amer. Soc. Civil Engrs* **103**, GT7, 757–768.

Seed, H.B. and Idriss, I.M. (1967) Analysis of soil liquefaction: Niigata earthquake. *Proc. Amer. Soc. Civil Engrs* **93**, SM3, 83–108.

Seed, H.B., Idriss, I.M. and Arango, I. (1983) Evaluation of liquefaction potential using field performance data. *J. Geotech. Engg ASCE* **109**, 458–482.

Seed, H.B., Martin, P.P. and Lysmer, J. (1976) Porewater pressure changes during soil liquefaction. *Proc. Amer. Soc. Civil Engrs* **102**, GT4, 323–344.

Sendlein, L.V.A., Yazicigil, H., Carlson, C.L. and Russell, H.K. (eds.) (1983) *Surface Mining, Environmental Monitoring and Reclamation Handbook*. Elsevier, New York, 750pp.

Shadbolt, C.H. (1975) Mining subsidence. In Bell, F.G. (1975a) *op. cit.*, 109–124.

Shadbolt, C.H. (1978) Mining subsidence—historical review and state of the art. In Geddes, J.D. (1978) *op. cit.*, 705–748.

Sheorey, P.R., Das, M.N., Barat, D., Prasad, R.K. and Singh, B. (1987) Coal pillar strength estimation from failed and stable cases. *Int. J. Rock Mechanics Ming Sci.* **24**, 347–355.

Shih, S.F., Mishoe, J.W., Jones, J.W. and Myhre, D.L. (1979) Subsidence related to land use in Everglades Agricultural Area. *Trans. Amer. Soc. Agric. Eng.* **22**, 560–568.

Shoham, D. and Levin, I. (1968) Subsidence on the reclaimed Hula Swamp area of Israel. *Israel J. Agric. Res.* **18**, 15–18.

Simons, N.E. (1975) Normally consolidated and lightly over-consolidated cohesive materials. In *Settlement of Structures*, British Geotechnical Society and Pentech, London 500–530.

Simons, N.E. (1987) Settlement. In Bell, F.G. (1987a) *op. cit.*, 14/1–14/27.

Sinclair, W.C. (1982) Sinkhole development resulting from groundwater withdrawal in the Tampa area, Florida. US Geological Survey Water Resources Investigation **81–50**, 19pp.

Singh, M.M. (1979) Experience with subsidence due to mining. In Saxena, S.K. (1979) *op. cit.*, 92–112.

Skempton, A.W. (1970) The consolidation of clays by gravitational compaction. *Q. J. Geol. Soc. London* **125**, 373–411.

Skempton, A.W. and MacDonald, D.H. (1956) Allowable settlement of buildings. *Proc. Instn Civil Engrs, Part 3*, **5**, 727–768.

Sladen, J.A., Bodimeade, C.S. and Jobling, V.R. (1984) Site investigation and urban development guidelines with respect to mining subsidence hazards—two examples from Alberta, Canada. In Forde, M.C. *et al.* (1984) *op. cit.*, 196–220.

Slowikowski, L. (1978) Contribution to discussion. In Geddes, J.D. (1978) *op. cit.*, 1054–1057.

Smedley, N. (1977) Subsidence management in the North Derbyshire area. *Ming Eng. London* **136**, 185–192.

Smith, D.L. and Smith G.L. (1987) Use of vertical gravity gradient analyses to detect near-surface dissolution voids in karst terrains. In Beck, B.F. and Wilson, W.L. (1987) *op. cit.*, 205–209.

Snowden, J.O., (1986) Drainage-induced land subsidence in metropolitan New Orleans, Louisiana, USA. International Association Hydrological Sciences Publ. **151**, 507–527.

Snowden, J.O., Simmons, W.B., Traughber, E.B. and Stephens, R.W. (1977) Differential subsidence of marshland peat as a geologic hazard in the Greater New Orleans area, Louisiana. *Trans. Gulf Coast Assn Geol. Soc.* **27**, 169–179.

Snyder, G.H., Burdine, H.W., Crockett, J.R., Gascho, G.J., Harrison, D.S., Kidder, G., Mishoe, J.W., Myhre, D.L., Pate, F.M. and Shih, S.F. (1978) Water-table management for organic soil conservation and crop production in the Florida Everglades. Florida Agricultural Experimental Station Bulletin **801**, 22pp.

Soderberg, A.D. (1979) Expect the unexpected: foundations for dams in karst. *Bull. Assoc. Engg Geol.* **16**, 409–425.

Song, L.H. (1986) Pumping subsidence of ground surface in karst areas. Academia Sinica, Beijing, 15pp.

Sotiropoulos, E. and Cavounidis, S. (1979) Cast in-situ piles in carstic limestone. *Proc. Conf. on Recent Developments in the Design and Construction of Piles*, Institute of Civil Engineers, London, 59–66.

Sowers, G.F. (1975) Failures in limestones in humid subtropics. *Proc. Amer. Soc. Civil Engrs* **101**, GT8, 771–787.

Sowers, G.F. (1984) Correction and protection in limestone terrane. In Beck, B.F. (1984) *op. cit.*, 373–378; also in *Envir. Geol. Water Sci.* (1986) **8**, 77–82.

Speck, R.C. and Bruhn, R.W. (1983) The Appalachian field: a surface monitoring program over pillar-extraction mine panels. In Sendlein, L.V.A. *et al.* (1983) *op. cit.*, 647–656.

Spencer, C.B. (1953) Leaning Tower of Pisa. *Engg News Record* **150**, 40–43.

Sperling, C.H.B., Goudie, A.S., Stoddart, D.R. and Poole, G.G. (1977) Dolines of the Dorset chalklands and other areas in southern Britain. *Trans. Inst. Brit. Geographers* **2**, 205–223.

Stacey, T.R. (1986) Interaction of underground mining and surface development in a central city environment. *Trans. Inst. Ming. Metall.* **95**A, 176–180.

Statham, I. and Baker, M. (1986) Foundation problems on limestone: a case history from the Carboniferous Limestone at Chepstow, Gwent, *Q. J. Engg Geol.* **19**, 191–201.

Statham, I., Golightly, C. and Treharne, G. (1987) Thematic mapping of the abandoned mining hazard: a pilot study for the South Wales coalfield. In Culshaw, M.G. *et al.* (1987) *op. cit.*, 255–268.

Stephens, J.C. (1958) Subsidence of organic soils in the Florida Everglades. *Proc. Soil Sci. Soc. Amer.* **20**, 77–80.

Stephens, J.C. (1974) Subsidence of organic soils in the Florida Everglades—a review and update. *Mem. Miami Geol. Soc.* **2**, 352–361.

Stephens, J.C., Allen, L.H. and Chen, E. (1984) Organic soil subsidence. *Geol. Soc. Amer. Rev. Engg Geol.* **6**, 107–122.

Stephens, J.C. and Stewart, E.H. (1977) Effect of climate on organic soil subsidence. International Association Hydrological Sciences Publ. **121**, 647–655.

Stephenson, R.W. and Aughenbaugh, N.B. (1978) Analysis and prediction of ground subsidence due to coal mine entry collapse. In Geddes, J.D. (1978) *op. cit.*, 100–118.

Stump, D., Nieto, A.S. and Ege, J.R. (1982) An alternative hypothesis for sink development above solution-mine cavities in the Detroit area. US Geological Survey Open-File Report, **82–297**, 61pp.

Su, H. (1986) Mechanism of land subsidence and deformation of soil layers in Shanghai. International Association Hydrological Sciences Publ. **151**, 425–433.

Subsidence Compensation Review Committee (1984) *The repair and compensation system for coal mining subsidence damage*. Department of Energy, London, 98pp.

Swiger, W.F. and Estes, H.M. (1959) Major power station foundation in broken limestone. *Proc. Amer. Soc. Civil Engrs* **85**, SM5, 77–86.

Symons, M.V. (1978) Sources of information for preliminary site investigation in old coal mining areas. In Geddes, J.D. (1978) *op. cit.*, 119–135.

Symons, M.V. (1985) Site investigation in old coal mining areas—recommended procedure for the desk study. In Geddes, J.D. (1985) *op. cit.*, 173–187.

Tan, B.K. (1987) Some geotechnical aspects of urban development over limestone terrain in Malaysia. *Bull. Int. Assoc. Engg Geol.* **35**, 57–63.

Tan, B.K. and Batchelor, B. (1981) Foundation problems in limestone areas: a case study in Kuala Lumpur, Malaysia. *Proc. Int. Symp. on Weak Rock, Tokyo*, Balkema, Rotterdam, 1461–1463.

Tandanand, S. and Powell, L.R. (1984) Influence of lithology on longwall mining subsidence. *Ming Engg* **36**, 1666–1671.

Taylor, R.K. (1975) Characteristics of shallow coal mine workings and their implications in urban redevelopment area. In Bell, F.G. (1975a) *op. cit.*, 125–148.

Terracina, F. (1962) Foundations of the Tower of Pisa. *Géotechnique* **12**, 336–339.

Thomas, G.G. (1987) A cost-benefit analysis for stabilising shallow Bath Stone mine workings at Corsham, Wiltshire. In Bell, F.G. *et al.* (1987) *op. cit.*, 157–165.

Thomas, T.M. (1974) The South Wales interstratal karst. *Trans. Brit. Cave Res. Assn* **1**, 131–152.

Thorburn, S. and Reid, W.M. (1978) Incipient failure and demolition of two storey dwellings due to large ground movements. In Geddes, J.D. (1978) *op. cit.*, 87–99.

Thornley, J.H., Spencer, C.B. and Albin, P. (1955) Mexico's Palace of Fine Arts settles 10 feet. *Civil Engg* **25**, 356–360.

Tinsley, J.C., Youd, T.L., Perkins, D.M. and Chen, A.T.F. (1985) Evaluating liquefaction potential. US Geological Survey Prof. Pap. **1360**, 263–315.

Tokimatsu, K. and Seed, H.B. (1987) Evaluation of settlements in sands due to earthquake shaking. *J. Geotech. Engg ASCE* **113**, 861–878.

Toms, A.H. (1966) Chalk in cuttings and embankments. *Proc. Symp. on Chalk in Earthworks and Foundations*. Institution of Civil Engineers, London, 43–54 and 90–95.

Tourtelot, H.A. (1974) Geologic origin and distribution of swelling clays. *Bull. Assn Engg Geol.* **11**, 259–275.

Twidale, C.R. (1987) Sinkhole (dolines) in lateritised sediment, Western Sturt Plateau, Northern Territory, Australia. *Geomorphology* **1**, 33–52.

Upchurch, S.B. and Littlefield, J.R. (1987) Evaluation of data for sinkhole development risk models. In Beck, B.F. and Wilson, W.L. (1987) *op. cit.*, 359–364.

Veder, H.C. (1976) The Leaning Tower of Pisa—a suggested remedial measure. *Ground Engg* **9**(1), 38–40.

Venter, I.S. and Gregory, B.J. (1987) Risk assessment in dolomitic terrain: a case history. In Culshaw, M.G. *et al.* (1987) *op. cit.*, 329–334.

von Post, L. (1922) Swedish Geological Survey peat inventory and some preliminary results (in Swedish). *Svensk. Mosskulturføreningens Tidskr.* **36**, 1–27.

Vos, G., Claessen, F.A.M and van Ommen, J.H.G. (1986) Geohydrological compensatory measures to prevent land subsidence as a result of the reclamation of the Markerwaard polder in the Netherlands. International Association Hydrological Sciences Publ. **151**, 915–928.

Wagener, F.V.M. and Day, P.W. (1986) Construction on dolomite in South Africa. *Envir. Geol. Water Sci.* **8**, 83–89.

Wallwork, K.L. (1960) Some problems of subsidence and land use in the Mid-Cheshire industrial area. *Geog. J.* **126**, 191–199.

Walters, R.F. (1977) Land subsidence in central Kansas related to salt dissolution. *Bull. Kansas Geol. Surv.* **214**, 82pp.

Waltham, A.C. (1986) Steel site starts Shuicheng sinkholes. *New Civil Engr* **672**, 24–25.

Waltham, A.C. and Smart, P.L. (1988) Civil engineering difficulties in the Karst of China. *Q. J. Engg Geol.* **21**, 2–6.

Waltham, A.C., Vandenven, G. and Ek, C.M. (1986) Site investigations on cavernous limestone for the Remouchamps Viaduct, Belgium. *Ground Engg* **19**(8), 16–18.

Waltham, T. (A.C.) (1978) *Catastrophe: the Violent Earth*. Macmillan, London, 170pp.

Ward, T. (1900) The subsidence in and around the town of Northwich in Cheshire. *Trans. Inst. Ming Engrs* **19**, 241–264.

Ward, W.H., Burland, J.H. and Gallois, R.W. (1968) Geotechnical assessment of a site at Mundsford, Norfolk, for a large proton accelerator. *Géotechnique* **18**, 399–431.

Wardell, K. (1957) The minimisation of surface damage. *Colliery Engg* **34**, 361–367.

Wassmann, T.H. (1979) Mining subsidence in the east Netherlands. In Saxena, S.K. (1979) *op. cit.*, 283–302.

Wegrzyn, M., Soto, A.E. and Perez, J.A. (1984) Sinkhole development in north-central Puerto Rico. In Beck, B.F. (1984) *op. cit.*, 137–142.

Weir, W.W. (1950) Subsidence of peat lands of the Sacramento San Joaquin delta, California. *Hilgardia* **20**(3), 37–56.

West, G. and Dumbleton, M.J. (1972) Some observations on swallow holes and mines in the chalk. *Q. J. Engg Geol.* **5**, 171–177.

Westbrook G.K., Kusznir, N.J., Browitt, C.W.A. and Holdsworth, B.K. (1980) Seismicity induced by coal mining in Stoke-on-Trent (UK). *Engg Geol.* **16**, 225–241.

Whitcut, M. (1981) Derelict land and reclamation problems in Telford. *Chartered Land Min. Surv.* **3**(1). 3–26.

White, E.L., Aron, G. and White, W.B. (1984) The influence of urbanization on sinkhole development in central Pennsylvania. In Beck, B.F. (1984) *op. cit.*, 275–281.

White, L.S. (1953) Transcona elevator failure: eye-witness account. *Géotechnique* **3**, 209–214.

White, W.B. (1984) Rate processes: chemical kinetics and karst landform development. In LaFleur, R.G. (ed.) *Groundwater as a Geomorphic Agent*, Allen and Unwin, London, 227–248.

Willis, A.J. and Chapman, J.A.F. (1980) Old coal workings—client, consultant and contractor. *Ground Engg* **13**(10), 22–39.

Wilson, G. and Grace, H. (1942) The settlement of London due to underdrainage of the London Clay. *Proc. Instn Civil Engrs* **2**, 100–116.

Wilson, J.G., Garwood, T.G. and Sarsby, R.W. (1985) The settlement of low-rise buldings constructed over peat. In Geddes, J.D. (1985) *op. cit.*, 526–538.

Wilson, S.D. (1987) A discussion of the Baldwin Hills Reservoir failure. *Engg Geol.* **24**, 127–141.

Winfield, P.F. (1984) Foundations for sites over natural voids and old mine workings. In Forde, M.C. *et al.* (1984) *op. cit.*, 266–272.

Wood, C. (1981) Exploration and geology of some lava tube caves on the Hawaiian volcanoes. *Trans. Brit. Cave Res. Assn* **8**, 111–129.

Wood, C.C. and Renfrey, G.J. (1975) The influence of mining subsidence on urban development of Ipswich, Queensland. *Proc. Second Australia and New Zealand Conf. on Geomechanics, Brisbane*, Institute of Engineers of Australia Publication **75/4**, 4–9.

Working Party of Geological Society (1982) Land surface evaluation for engineering practice. *Q. J. Engg Geol.* **15**, 265–316.

Yamamoto, S. (1984) Case history no. 9.4. Tokyo, Japan. In Poland, J.F. (1984) *op. cit.*, 175–184.

Yoshimi, Y. (1980) Protection of structures from soil liquefaction hazards. *Geotech. Engg* **11**, 181–208.

Youd, T.L. and Perkins, D.M. (1978) Mapping liquefaction-induced ground failure potential. *Proc. Amer. Soc. Civil Engrs* **104**, GT4, 433–446.

Youd, T.L. and Perkins, D.M. (1987) Mapping of liquefaction severity index. *J. Geotech. Engg* **113**, 1374–1392.

Young, N.M. (1970) Caves in southeast England. *J. Brit. Speleological Assn* **45**, 13–20.

Yuan, D. (1983) *Problems of environmental protection of karst area*. Institute of Karst Geology, Guilin, 15pp.

Yuan, D. (1987) Environmental and engineering problems of karst geology in China. In Beck, B.F. and Wilson, W.L. (1987) *op. cit.*, 1–11.

Zeevaert, L. (1957) Foundation design and behaviour of the Tower Latino Americana in Mexico City. *Géotechnique* **7**, 115–133.

Zeevaert, L. (1972) *Foundation Engineering for Difficult Subsoil Conditions*. Van Nostrand, New York, 652pp.

Zeevaert, L. (1987) Design of compensated foundations. In Bell, F.G. (1987) *op. cit.*, 51/1–51/20.

Zhang, S. (1984) Karst and subsidence in China. In Beck, B.F. (1984) *op. cit.*, 97–104.

Index